Advanced Therapy in Companion Animals—2nd Edition

Advanced Therapy in Companion Animals—2nd Edition

Guest Editors

Zhijun Zhong
Ziyao Zhou

MDPI

Basel • Beijing • Wuhan • Barcelona • Belgrade • Novi Sad • Cluj • Manchester

Guest Editors
Zhijun Zhong
Department of Veterinary Surgery
Sichuan Agricultural University
Chengdu
China

Ziyao Zhou
Department of Veterinary Surgery
Sichuan Agricultural University
Chengdu
China

Editorial Office
MDPI AG
Grosspeteranlage 5
4052 Basel, Switzerland

This is a reprint of the Special Issue, published open access by the journal *Veterinary Sciences* (ISSN 2306-7381), freely accessible at: https://www.mdpi.com/journal/vetsci/special_issues/5AEK176G58.

For citation purposes, cite each article independently as indicated on the article page online and as indicated below:

Lastname, A.A.; Lastname, B.B. Article Title. *Journal Name* **Year**, *Volume Number*, Page Range.

ISBN 978-3-7258-6562-8 (Hbk)
ISBN 978-3-7258-6563-5 (PDF)
https://doi.org/10.3390/books978-3-7258-6563-5

Contents

veterinary sciences

MDPI

Article

Identification and Characterization of Canine Parvoviruses and Emergence of Canine Bocavirus and Bufavirus from Diarrheic Dogs in Sichuan Province, China

Siyu Liu †, Xiaoqi Li †, Yuxin Zhou †, Shuangshuang Song, Yuyan Huang, Mengjie Che, Xin Lei, Iram Laghari, Mingyue Wu, Ruilin Han, Haifeng Liu, Ziyao Zhou, Guangneng Peng, Kun Zhang and Zhijun Zhong *

Key Laboratory of Animal Disease and Human Health of Sichuan, College of Veterinary Medicine, Sichuan Agricultural University, Chengdu 611130, China; llsy6412024@163.com (S.L.)

* Correspondence: zhongzhijun488@126.com

† These authors contributed equally to this work.

Simple Summary

Canine diarrhea is frequently associated with viral infections. While canine parvovirus (CPV) is a well-known pathogen, two recently discovered viruses—canine bocavirus (CBoV) and canine bufavirus (CBuV)—are also recognized as potential contributors. This study investigated the presence and characteristics of these three viruses in diarrheic dogs from five regions in Sichuan, China. Our findings showed that CPV-2 (33.3%, 48/144) was the most prevalent virus. Notably, this is the first report of CBoV (5.56%, 8/144) and CBuV (4.17%, 6/144) detection in Sichuan. The dominant CPV strain identified was CPV-2c, which had replaced previously circulating strains such as CPV-2a and new CPV-2a. We also identified novel genetic mutations and documented several co-infection cases where dogs were infected with multiple viruses. These results reveal a dynamic and evolving viral situation in Sichuan's dog population, highlighting the importance of ongoing surveillance to effectively manage disease spread and protect canine health.

Abstract

Canine parvovirus (CPV) is the primary cause of viral enteritis in dogs, while canine bocavirus (CBoV) and canine bufavirus (CBuV) have emerged as significant components of the canine enteric virome. Previous studies on CPV in Sichuan had limited geographic coverage and sample size, and lacked data on CBoV and CBuV. This study used PCR to investigate the genetic diversity of CPV, CBoV and CBuV and to examine their co-infection status in diarrheic dogs across five regions of Sichuan between 2020 and 2022. The results revealed that CPV-2 was the most prevalent virus (33.3%, 48/144), while CBoV (5.56%, 8/144) and CBuV (4.17%, 6/144) were detected for the first time in Sichuan province. Genetic analysis revealed CPV-2c to be the predominant genotype (95.8% (46/48)), replacing the previously circulating strains CPV-2a and CPV-new 2a. The study also identified several typical and novel mutations in the VP2 protein in the CPV-2c strains, including Ala5Gly, Trp214Cys and Thr440Ala. Additionally, four co-infection cases (2.78%) were observed, including a triple CPV/CBoV/CBuV infection in Xichang (XC05). Phylogenetic analysis revealed genetic diversity, with CBoV strains clustering into CBoV-1 and CBoV-2 subspecies, while CBuV strains formed two distinct clusters. Our findings emphasize the need for ongoing monitoring of the dynamic epidemiological situation of CPV, CBoV and CBuV in Sichuan Province.

Keywords: canine parvovirus; canine bocavirus; canine bufavirus; VP2; phylogenetic analysis

1. Introduction

Canine infectious diarrhea poses a considerable challenge to veterinary practitioners due to the multifaceted etiology of the condition, which encompasses viral infections, parasitic infections and gut microbiota dysbiosis [1–3]. Among the viral pathogens, canine parvovirus (CPV), canine bocavirus (CBoV) and canine bufavirus (CBuV), all belonging to the genus Protoparvovirus, are three major enteric viruses of concern. CPV-2 has been identified as the primary causative agent of severe viral diarrhea, particularly in puppies [4]. CBoV and CBuV are newly discovered parvoviruses that have also been associated with canine diarrhea and require further investigation into their epidemiology [5,6]. Co-infections involving the three viruses are commonly observed and may exacerbate disease progression, leading to more severe clinical manifestations [7,8].

CPV is a single-stranded DNA virus that contains two large open reading frames (ORFs). It has been established that ORF1 encodes two non-structural proteins, designated as NS1 and NS2, while ORF2 encodes two structural proteins, termed VP1 and VP2 [9,10]. VP2, the major capsid protein of CPV, is the primary determining factor in genotype classification, based on variations at key antigenic sites, notably residues 426 and 297 [11]. The continuous molecular evolution of the key VP2 residues has driven the emergence and dominance of new antigenic variants in China. Three antigenic variants—CPV-2a (426Asn), CPV-2b (426Asp), and CPV-2c (426Glu)—are determined by variations at residue 426 of the VP2, while novel strains (e.g., new CPV-2a/2b) are distinguished by Ser297Ala [12–14]. To date, all major CPV-2 genotypes (including CPV-2a, -2b, -2c, and new variants) have been identified across China, with the dominant CPV genotypes exhibiting variation across Chinese provinces. It has been reported that CPV-2c is the predominant strain in Shanghai, Shandong, Tangshan, Jilin, Guangxi, and Henan provinces [15–21]. Meanwhile, CPV-2a and new CPV-2a have shown prominence in regions including Gansu, as well as in Heilongjiang and Jiangsu, respectively [22–24]. In Sichuan, only four studies have focused on the molecular epidemiology of CPV, and these studies are constrained by several limitations, including small sample sizes (e.g., samples number less than 58) and a restricted geographical focus (often limited to Chengdu, the capital of Sichuan Province) [25–28].

CBoV and CBuV are recently discovered parvoviruses in China that are associated with gastroenteritis of varying severity in canines [29,30]. Of the three genotypes of CBoV (CBoV-1, -2, and -3) that have been reported on a global scale, CBoV-2 has been confirmed as enteropathogenic [31–35]. In China, CBoV has been reported in Heilongjiang [36], Guangzhou [31] and Hong Kong [37]. The first documented detection of CBuV was in dogs with intestinal and respiratory tract disease in Italy in 2018 [38]. Subsequent reports have emerged from China [39,40], India [41] and Turkey [42]. In China, CBuV was first documented in Guangxi in 2019 [39], with subsequent reports from Shanghai, Henan and Anhui [40,43,44]. It is important to note that co-infection with CBuV and CPV have been observed in several studies [39,45]. Furthermore, these viruses have also been identified from apparently healthy animals [46]. However, further research is required to elucidate the molecular epidemiology of CBoV and CBuV in Sichuan Province.

According to the 2025 China Pet Industry White Paper, the number of pet dogs in China was 52.58 million in 2024 [47]. Sichuan, a province with a substantial population, is characterized by a significant canine population. At present, molecular epidemiological data and co-infection patterns on CBoV, CBuV and CPV remain limited and require updat-

https://doi.org/10.3390/vetsci13010041

ing in Sichuan Province. To address these knowledge gaps, the aim of the present study was to investigate the characteristics of CPV, CBoV and CBuV in Sichuan province, and also investigated their co-infection status.

2. Materials and Methods

A total of 144 fecal samples were collected from 144 dogs that presented with diarrhea and accompanying clinical signs (including fever, vomiting, anorexia, or lethargy) to local veterinary clinics for clinical care. All the samples were collected from five districts in Sichuan Province between November 2020 and June 2022. The clinical symptoms and vaccination history of each dog were recorded at the time of sampling. Samples were collected with an anal swab and placed into a 1.5 mL centrifuge tube containing 1.0 mL of phosphate-buffered saline (PBS). The samples were obtained through convenience sampling of clinically affected dogs presented for veterinary care. The samples were sourced from five distinct regions with a high density of domestic dogs: Chengdu (n = 44), Ya'an (n = 35), Mianyang (n = 21), Leshan (n = 29), and Xichang (n = 15). All samples were stored at −20 °C until further processing.

Total viral DNA was extracted from fecal samples using the QIAamp PowerFecal Pro DNA Kit (Qiagen Co., Ltd., Hilden, Germany) in accordance with the manufacturer's protocols. The extracted DNA was then amplified using polymerase chain reaction (PCR) with previously published primer sets targeting specific genes: the complete VP2 gene of CPV-2 (1755 bp), the NS1 gene of CBoV (440 bp), and the VP2 gene of CBuV (962 bp) [23,25,48]. The primer sequences used are detailed in Table S1. The reactions were performed in a total volume of 50 μL (for CPV and CBoV) or 25 μL (for CBuV) under three different reaction conditions (Table S1). The amplification products of the genes were sent to Sangon Biotech (Shanghai) Co., Ltd., Shanghai, China for bidirectional sequencing analysis.

The obtained gene sequences (the complete VP2 gene of CPV, the NS1 gene of CBoV and the VP2 gene of CBuV) were initially analyzed using the BLAST web interface (National Center for Biotechnology Information, Bethesda, MD, USA; https://blast.ncbi.nlm.nih.gov/Blast.cgi, accessed on 15 May 2024) to identify and download the reference sequences (95 strains of CPV, 16 strains of CBoV, and 23 strains of CBuV) from the GenBank database. Multiple sequence alignments for each virus were performed using MEGA software (version 11.0; Mega Limited, Auckland, New Zealand). Specifically, the phylogenetic tree for CPV was reconstructed using the Maximum Likelihood method based on the Tamura 3-parameter model with gamma distribution, and the trees for CBoV and CBuV were constructed using the Neighbor-Joining method with the Kimura 2-parameter model. The robustness of the nodes was tested using 1000 bootstrap replications and bootstrap support values greater than 60 were shown at the nodes. Additionally, the VP2 genes of CPV and CBuV, along with the NS1 gene of CBoV, were translated into amino acid sequences using MEGA (version 11.0). The nucleotide and amino acid sequence identities between the isolates from this study and the reference strains included in the phylogenetic trees were calculated using the MegAlign module within the BioEdit software (version 7.2.5; Ibis Biosciences, Carlsbad, CA, USA).

The calculation of detection rates and the subsequent determination of 95% confidence intervals (95% CI) was conducted utilizing the statistical software SPSS 22.0 for Windows (SPSS Inc., Chicago, IL, USA).

3. Results

3.1. Distribution of CPV, CBoV, and CBuV in Diarrheic Dogs

Among the 144 fecal samples collected from diarrheic dogs across five cities in Sichuan Province between 2020 and 2022, CPV-2 was the most frequently detected virus (33.3%, 48/144; 95% CI: 25.6–41.0%), followed by CBoV (5.56%, 8/144; 95% CI: 1.8–9.3%)

https://doi.org/10.3390/vetsci13010041

and CBuV (4.17%, 6/144; 95% CI: 0.9–7.4%) (Figure 1, Figure 2 and Figure S1). CPV-2 and CBoV were detected in all five regions, whereas CBuV was only detected in Chengdu (4.55%, 2/44; 95% CI: 0.0–10.7%), Ya'an (5.71%, 2/35; 95% CI: 0.0–13.4%), and Xichang (13.33%, 2/15; 95% CI: 0.0–30.5%). The highest detection rate of CPV-2 was observed in Mianyang (61.9%, 13/21; 95% CI: 41.1–82.7%), whereas the lowest was detected in Leshan (24.13%, 7/29; 95% CI: 8.6–39.7%). For CBoV, the highest detection rate was observed in Leshan (6.7%, 2/29; 95% CI: 0.0–15.8%), while the lowest was detected in Ya'an (2.86%, 1/35; 95% CI: 0.0–8.4%).

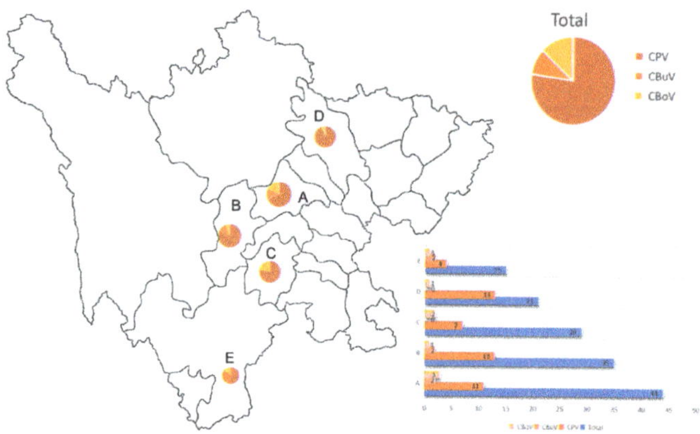

Figure 1. Geographical locations and sample statistics of the isolates analyzed in this study with positive rates of three viruses. In the histogram, the various color bars show the number of sequenced samples. The pie chart illustrates the proportion of each virus detected in the total number of samples analyzed. Note. 1: A: Chengdu city; B: Ya'an city; C: Leshan city; D: Mianyang city; E: Xichang city.

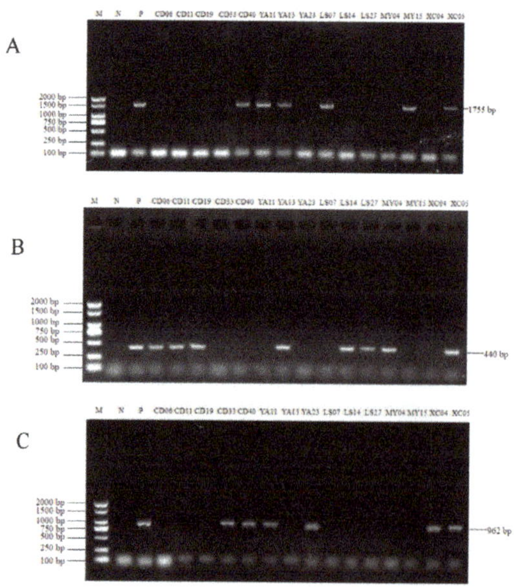

Figure 2. (**A**) Specific fragment of the CPV-2 VP2 gene (1755 bp) was amplified with CPV-VP2's primer. (**B**) Specific fragment of the CBoV NS1 gene (440 bp) was amplified with CBoV-NS1's primer. (**C**) Specific fragment of the CBuV VP2 gene (962 bp) was amplified with CBuV-VP2's primer. Lane M, TaKaRa DL 2000 DNA Marker; lane N, negative control; lane P, positive control.

https://doi.org/10.3390/vetsci13010041

Clinical manifestations among virus-positive dogs are summarized in Table 1. Among the 48 CPV-2–positive dogs, 10 dogs exhibited hemorrhagic diarrhea, accounting for 20.8%. In contrast, none of the eight CBoV-positive dogs presented with hemorrhagic diarrhea. Among the six CBuV-positive dogs, only one case (16.7%) was observed with hemorrhagic diarrhea. The vaccination status of virus-positive dogs was also analyzed. Among the 48 CPV-2–positive dogs, 44 (91.67%) had either not received vaccination or had not completed the full vaccination course. Similarly, incomplete or absent vaccination was observed in six of the eight CBoV-positive dogs (75.0%) and in all six CBuV-positive dogs (100%). Based on individual-level clinical and virological data, four cases of co-infection were also identified (Table 1): one case of CPV/CBuV co-infection in Chengdu (CD40); one case of CPV/CBuV and one case of CPV/CBoV co-infection in Ya'an (YA11/13); and one triple CPV/CBoV/CBuV co-infection in Xichang (XC05). All co-infected dogs were young (<1 year) and presented with a consistent set of clinical signs, including diarrhea, lethargy, and anorexia.

Table 1. Clinical information and infection status of virus positive samples.

Number	Source	Breed	AGE	Clinical Symptoms	Vacccination	Infected Virus			Genbank Number
						CPV	CBoV	CBuV	
CD03	Chengdu	Pomeranian	3 months	■◆▲ (blue, green, grey)	No	+	-	-	CPV:MZ857180
CD04	Chengdu	Chinese Field Dog	5 months	●■▲ (red, yellow, grey)	No	+	-	-	CPV:MZ857181
CD05	Chengdu	Chinese Field Dog	2 months	●■▲ (red, yellow, grey)	No	+	-	-	CPV:MZ857185
CD06	Chengdu	Poodle	7 years	■◆ (yellow, green)	Complete	-	+	-	CBoV:OQ030256
CD08	Chengdu	Chinese Field Dog	4 months	■◆▲ (yellow, green, grey)	Incomplete	+	-	-	CPV:OP779632
CD09	Chengdu	Poodle	3 months	■◆▲ (blue, green, grey)	No	+	-	-	CPV:OP779633
CD10	Chengdu	Chinese Field Dog	2 months	■▲ (yellow, grey)	No	+	-	-	CPV:OP779634
CD11	Chengdu	Chinese Field Dog	4 months	■◆ (yellow, green)	No	-	+	-	CBoV:OQ030257
CD14	Chengdu	Chinese Field Dog	16 months	■◆▲ (yellow, green, grey)	No	+	-	-	CPV:OP779635
CD19	Chengdu	Chinese Field Dog	1.5 months	■▲ (yellow, grey)	No	-	+	-	CBoV:OQ030258
CD21	Chengdu	Chinese Field Dog	2 months	●■▲ (red, yellow, grey)	No	+	-	-	CPV:OP779636
CD30	Chengdu	Poodle	-	■◆ (yellow, green)	No	+	-	-	CPV:OP779637
CD33	Chengdu	Chinese Field Dog	2 months	■◆▲⬟ (blue, green, grey, purple)	No	-	-	+	CBuV:OQ030264
CD38	Chengdu	Pomeranian	2 months	●■◆▲ (red, orange, green, grey)	No	+	-	-	CPV:OP779638
CD40	Chengdu	Welsh Corgi	2 months	●■◆▲⬟ (red, orange, green, grey, purple)	No	+	-	+	CPV:OP779639 CBuV:OQ030265
YA04	Ya'an	Golden Retriever	3 months	●■▲ (red, yellow, grey)	Incomplete	+	-	-	CPV:MZ857182
YA05	Ya'an	Welsh Corgi	2 months	●■◆▲ (red, yellow, green, grey)	Incomplete	+	-	-	CPV:MZ857183
YA06	Ya'an	Labrador Retriever	3 months	●■▲ (red, yellow, grey)	Incomplete	+	-	-	CPV:MZ857186

https://doi.org/10.3390/vetsci13010041

Table 1. *Cont.*

Number	Source	Breed	AGE	Clinical Symptoms	Vacccination	Infected Virus			Genbank Number
						CPV	CBoV	CBuV	
YA08	Ya'an	Chinese Field Dog	4 months	🟨 ◆	Incomplete	+	-	-	CPV:OP779640
YA10	Ya'an	Chinese Field Dog	3 months	🟨 ▲	Incomplete	+	-	-	CPV:OP779641
YA11	Ya'an	Greyhound	2 months	🟨 ◆ ▲	No	+	-	+	CPV:OP779642 CBuV:OQ030266
YA13	Ya'an	Chinese Field Dog	2 months	🟨 ◆ ▲	No	+	+	-	CPV:OP779643 CBoV:OQ030259
YA17	Ya'an	Chinese Field Dog	2 months	🔴 🟦 ◆ ▲	Incomplete	+	-	-	CPV:OP779644
YA18	Ya'an	Chinese Field Dog	9 months	🟨 ◆ ⬟	Incomplete	+	-	-	CPV:OP779645
YA23	Ya'an	Chinese Field Dog	3 monthes	🟦 ◆ ▲ ⬟	Incomplete	-	-	+	CBuV:OQ030267
YA25	Ya'an	Chinese Field Dog	2 months	🟨 ◆ ▲ ⬟	No	+	-	-	CPV:OP779646
YA27	Ya'an	Chinese Field Dog	2 months	🟦 ◆ ⬟	No	+	-	-	CPV:OP779647
YA29	Ya'an	Chinese Field Dog	2 months	🟦 ◆ ⬟	No	+	-	-	CPV:OP779648
YA32	Ya'an	Chinese Field Dog	2 months	🟧 ◆ ▲ ⬟	No	+	-	-	CPV:OP779649
LS03	Leshan	Golden Retriever	-	🔴 🟧 ⬟	No	+	-	-	CPV:OP779650
LS07	Leshan	Chinese Field Dog	5 months	🔴 🟧 ◆ ▲ ⬟	Incomplete	+	-	-	CPV:OP779651
LS08	Leshan	Chinese Field Dog	4 months	🔴 🟧 ◆ ▲ ⬟	Incomplete	+	-	-	CPV:OP779652
LS09	Leshan	Chinese Field Dog	4 months	🔴 🟧 ◆ ▲ ⬟	Incomplete	+	-	-	CPV:OP779653
LS14	Leshan	Welsh Corgi	2 years	🟦 ◆ ⬟	Complete	-	+	-	CBoV:OQ030260
LS20	Leshan	Poodle	3 months	🟧 ⬟	Incomplete	+	-	-	CPV:OP779654
LS21	Leshan	Golden Retriever	2 months	🟨 ⬟	Incomplete	+	-	-	CPV:OP779655
LS27	Leshan	Poodle	2 months	🟨 ◆ ▲	No	-	+	-	CBoV:OQ030261
LS29	Leshan	Chinese Field Dog	2 months	🟦 ◆ ▲ ⬟	No	+	-	-	CPV:OP779656
MY01	Mianyang	Chinese Field Dog	3 months	🔴 🟨 ▲	Incomplete	+	-	-	CPV:OP779657
MY02	Mianyang	Poodle	5 months	🟨 ▲	Incomplete	+	-	-	CPV:OP779658
MY04	Mianyang	Poodle	2 years	🟨 ◆ ▲	Incomplete	-	+	-	CBoV:OQ030262
MY05	Mianyang	Poodle	1 months	🔴 🟨 ▲	No	+	-	-	CPV:OP779659
MY06	Mianyang	Poodle	4 months	🟨 ▲	Incomplete	+	-	-	CPV:OP779660
MY07	Mianyang	Pomeranian	3 months	🟨 ◆ ▲	Incomplete	+	-	-	CPV:OP779661
MY09	Mianyang	Pug	3 months	🔴 🟨 ◆ ▲	Incomplete	+	-	-	CPV:OP779662
MY14	Mianyang	German Pinscher	2 months	🟨 ◆	No	+	-	-	CPV:OP779663
MY15	Mianyang	Greyhound	10 months	🔴 🟧 ◆ ▲ ⬟	Complete	+	-	-	CPV:OP779664
MY17	Mianyang	Chinese Field Dog	2 months	🟦 ◆ ▲ ⬟	No	+	-	-	CPV:OP779665

https://doi.org/10.3390/vetsci13010041

Table 1. *Cont.*

Number	Source	Breed	AGE	Clinical Symptoms	Vacccination	Infected Virus			Genbank Number
						CPV	CBoV	CBuV	
MY18	Mianyang	Chinese Field Dog	3 months	🟧 ◆ ▲	Incomplete	+	-	-	CPV:OP779666
MY19	Mianyang	Chinese Field Dog	2 months	🟦 ◆ ▲ ⬟	Incomplete	+	-	-	CPV:OP779667
MY20	Mianyang	Chinese Field Dog	3 months	🔴 🟦 ◆ ▲ ⬟	Incomplete	+	-	-	CPV:OP779668
MY21	Mianyang	Chinese Field Dog	6 months	🟨 ◆ ▲ ⬟	Complete	+	-	-	CPV:OP779669
XC02	Xichang	Golden Retriever	6 months	🔴 🟨 ◆ ▲	Complete	+	-	-	CPV:OP779670
XC04	Xichang	Poodle	4 months	🔴 🟨 ◆ ▲	Incomplete	-	-	+	CBuV:OQ030268
XC05	Xichang	Poodle	3 months	🟦 ◆ ▲	Incomplete	+	+	+	CPV:OP779671 CBoV:OQ030263 CBuV:OQ020369
XC06	Xichang	Welsh Corgi	3 months	🟨 ◆ ▲	Incomplete	+	-	-	CPV:OP779672
XC08	Xichang	Pomeranian	24 months	🔴 🟨 ◆ ▲	Complete	+	-	-	CPV:OP779673

Note. Fever 🔴; stool characteristics (bloody stool 🟧, watery stool 🟦, soft stool 🟨); anorexia ◆; lethargy ▲; vomiting ⬟.

3.2. Phylogenetic and Genetic Analysis of CPV-2

3.2.1. Amino Acid Analysis Suggests the Predominance of CPV-2c in Sichuan

Of the 48 CPV-2 strains analyzed, 46 (95.8%) were identified as CPV-2c (426-Glu), while two (4.2%) were identified as CPV-new 2a (297-Ala, 426-Asn). These results indicated that CPV-2c was the predominant genotype in Sichuan. The key AA sites for virus evolution were summarized in Table S2. All 46 CPV-2c strains in this study shared three characteristic amino acid substitutions (Phe267Tyr, Tyr324Ile and Gln370Arg) in VP2 with the original CPV-2 reference strain (M38245), which has been documented as a characteristic of CPV-2c strains of Asian origin. The majority (95.65%, 44/46) of the CPV-2c strains retained the traditional alanine (Ala) at VP2 position 5. However, two strains (CD21 and MY20) exhibited an Ala5Gly amino acid mutation. In addition to these typical VP2 amino acid mutations (Ala5Gly, Phe267Tyr, Tyr324Ile and Gln370Arg), several novel mutations were identified. These included a Met447Ile substitution in one CPV-2c strain (LS03), a Thr440Ala mutation in one CPV-2c strain (CD04) and in both CPV-new 2a strains (CD03 and YA06), and a Trp214Cys mutation in four CPV-2c strains from Xichang (XC02, XC05, XC06 and XC08).

3.2.2. VP2 Gene Phylogenetic Analysis Revealed Three Main CPV-2c Clades

The phylogenetic relationship of CPV-2 was reconstructed based on 143 complete VP2 gene sequences, including the 48 strains from this study. The phylogenetic tree showed that all sequences clustered into two major clades defined by key residues at VP2 positions 267 and 324. One clade was characterized by the amino acid substitutions Phe267Tyr/Tyr324Ile, predominantly found in Asian strains, while the other clade exhibited the residues Phe267/Tyr324, predominantly found in Western strains (Figure 3, Table S2). Five primary monophyletic groups were identified: Asia-2, Asia-1.1, Asia-1.2, WT-1 and WT-2. All CPV-2c strains from this study were grouped within the Asia-1.1 and Asia-1.2 subclades. In contrast, the two CPV-new 2a strains (CD03 and YA06) clustered within the Asia-2 subclade.

https://doi.org/10.3390/vetsci13010041

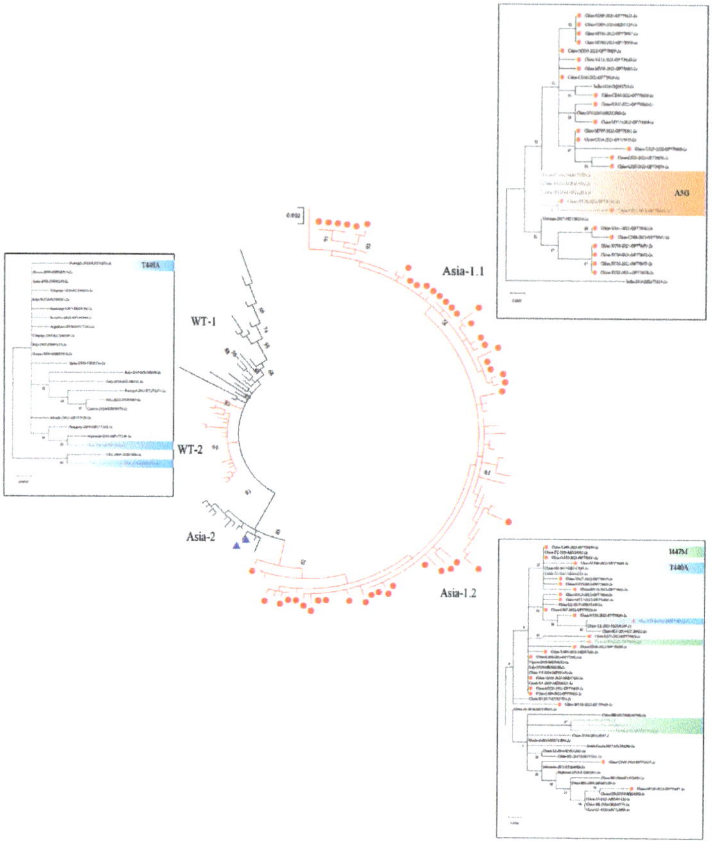

Figure 3. Phylogenetic tree of canine parvovirus (CPV) constructed based on the complete VP2 gene sequences. The tree was inferred using the Maximum Likelihood method with the Tamura 3-parameter model. A discrete Gamma distribution (5 categories, +G, parameter = 0.1000) was used to model evolutionary rate differences among sites. The tree is drawn to scale, with branch lengths measured in the number of substitutions per site. The analysis involved 138 nucleotide sequences, and there was a total of 1755 positions in the final dataset after the removal of all positions containing gaps and missing data (complete deletion option). Red circles indicate CPV-2c strains identified in the present study, and blue triangles indicate CPV-2a strains identified in this study.

The Asia-1.1 and Asia-1.2 subclades were differentiated by the residue at VP2 position 447 (Met/Ile). A total of 24 CPV-2c strains in this study were found to cluster within the Asia-1.1. The MY20 and CD21 strains showed a close phylogenetic relationship with three reference strains from China: two from Sichuan (MH476585 and MF805796) and one from Henan (ON322803). Notably, the amino acid sequence of the MY20 strain was 100% identical to that of these three reference strains. The Asia-1.2 subclade comprised CPV-2c strains from China (including 22 strains from this study), Vietnam, Thailand, South Korea, and Indonesia. Within the Asia-1.2 subclade, LS03 strain in our study and three Vietnamese strains (MW239584, MW239603 and MK357721) all carried the 447M residue but were located on different small branches. Another strain from this study (CD04) clustered with Chinese CPV-2c mutants (MZ836339 and KT156832). CD04 was found to carry the T440A mutation, which was also observed in two Western CPV-2c strains, one from Chile (MT585703) and one from the USA (FJ005236). In addition, four Xichang strains (XC02, XC05, XC06 and XC08) exhibited the Trp214Cys mutation and formed an

https://doi.org/10.3390/vetsci13010041

independent small branch on the evolutionary tree. The CPV-2c strains from this study shared 98.8-100% nucleotide identity and 98.6-100% amino acid identity among themselves (Tables S3 and S4). Notably, 54.35% (25/46) exhibited 100% amino acid identity with CPV-2c reference strains reported in several Chinese regions, Nigeria, and Italy from 2014 to 2021.

3.3. NS1 Gene Phylogenetic Analysis of CBoV

Phylogenetic analysis showed that the eight sequences from this study clustered into two distinct groups (Figure 4). Six of the strains were found within a major clade that contained 11 reference strains from Asia, America and Europe. The remaining two strains (OQ030256 and OQ030258) formed a well-supported subclade with four reference sequences from China (KR998493, KR998488 and KR998491) and Canada (OK546118). Based on their phylogenetic positions, all eight sequences were classified into two established CBoV subspecies: six strains belonged to CBoV2, and two strains (OQ030256 and OQ030258) were assigned to CBoV1. Notably, no strains from this study clustered with the CBoV3 subspecies.

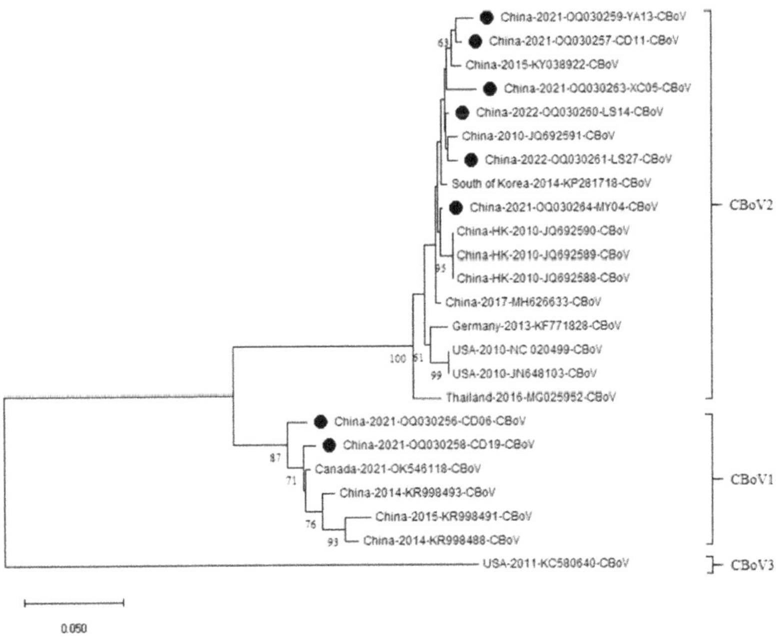

Figure 4. The evolutionary history was inferred using the Neighbor-Joining method. The optimal tree with the sum of branch length = 0.67599231 is shown. The percentage of replicate trees in which the associated taxa clustered together in the bootstrap test (1000 replicates) are shown next to the branches. The tree is drawn to scale, with branch lengths in the same units as those of the evolutionary distances used to infer the phylogenetic tree. The evolutionary distances were computed using the Kimura 2-parameter method and are in the units of the number of base substitutions per site. The analysis involved 24 nucleotide sequences. All positions containing gaps and missing data were eliminated. Black symbols indicate CBoV strains identified in the present study.

3.4. VP2 Gene Phylogenetic Analysis of CbuV

Phylogenetic analysis showed that the six sequences from this study clustered into two distinct genetic groups (Cluster 1 and Cluster 2, Figure 5). Cluster 1 was a well-supported major clade that included five CBuV strains from this study (CD33, CD40, XC04, XC05 and YA23) and reference sequences from China, Thailand, Italy and India. Within this major clade, the five CBuV strains from this study formed a distinct, monophyletic subcluster with moderate support. In the cluster 2, the remaining one strain (YA11) formed a well-supported subclade with four reference sequences from Italy, China and Turkey.

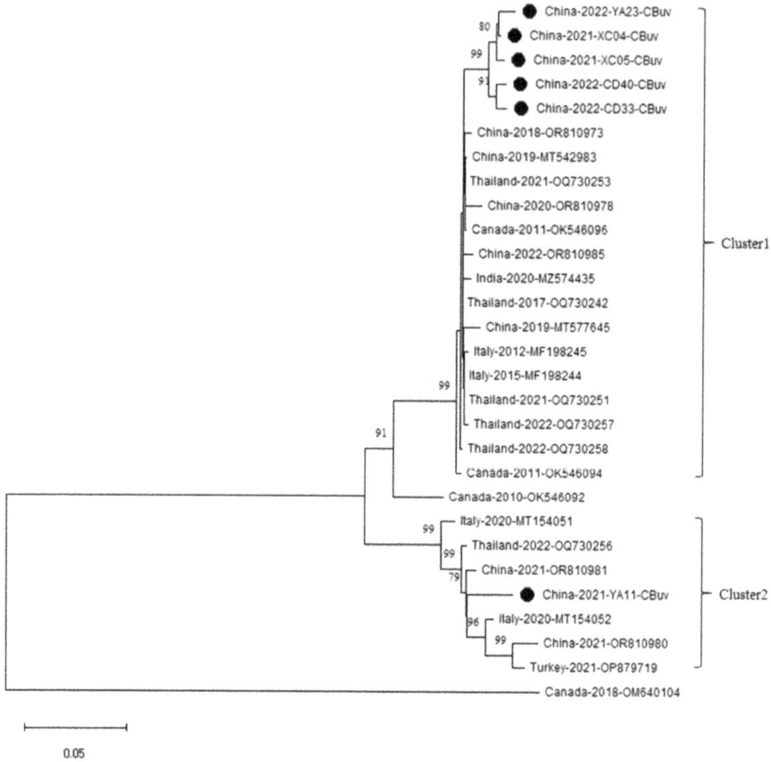

Figure 5. Phylogenetic analysis of CBuV based on VP2 gene sequences. The evolutionary history was inferred using the Neighbor-Joining method. The optimal tree with the sum of branch length = 0.73839789 is shown. The percentage of replicate trees in which the associated taxa clustered together in the bootstrap test (1000 replicates) are shown next to the branches. The tree is drawn to scale, with branch lengths in the same units as those of the evolutionary distances used to infer the phylogenetic tree. The evolutionary distances were computed using the Kimura 2-parameter method and are in the units of the number of base substitutions per site. The analysis involved 30 nucleotide sequences. All positions containing gaps and missing data were eliminated. Black symbols indicate CBuV strains identified in the present study.

4. Discussion

CPV is the primary cause of severe viral enteritis in dogs worldwide. In recent years, two novel parvoviruses (CBoV and CBuV) have been identified as recently discovered pathogens and are now widely recognized as common components of the canine enteric virome [5,6]. Notably, co-infections involving CPV, CBoV and CBuV have been reported, and these are considered to increase the complexity of viral enteritis [36,42]. Before this study, the epidemiology of CPV in Sichuan was limited by geographic coverage and sample size. Moreover, the presence and molecular characteristics of CBoV and CBuV in Sichuan had not been reported. This study examined 144 diarrheic dogs using PCR between 2020 and 2022 in Sichuan Province, exploring the characteristics of CPV, CBoV, and CBuV and analyzing their co-infection rates.

https://doi.org/10.3390/vetsci13010041

In the present study, we analyzed the detection rates and co-infection patterns of CPV-2, CBoV and CBuV in Sichuan. Our results showed CPV-2 was the most prevalent virus detected (33.3%) in diarrheic dogs across the five regions surveyed, indicating its status as a dominant and widespread enteric pathogen in Sichuan. The detection rates of CPV-2 in our study was lower than that reported in Shandong (62.9%), and Shanghai (40.78%), and East China (40.7%) [15,16,49]. Additionally, CBoV and CBuV were first detected in Sichuan, with positive rates of 5.56% (8/144) and 4.17% (6/144), respectively. Recent studies have indicated that CBoV has become endemic among dogs worldwide [5,7,31,33,35,37]. The detection rates of CBoV in our study was higher than that reported in Guangdong (0.24%) [31], but lower than in Heilongjiang (20%) [36]. It aligns closely with reports from Korea (9.6%) [33]. As to CBuV, only four surveys had analyzed the prevalence of CBuV in fecal samples from diarrheic in China before 2022. The detection rates of CBuV in our study (4.17%) was similar to that observed in Guangxi (6.25%) [39]. This is higher than in Henan (1.74%) and Anhui (4.4%) [43,44], but lower than that reported in Shanghai (42.51%) [40]. Furthermore, we identified four cases of co-infection (2.78%, 4/144), including one triple infection (CPV/CBoV/CBuV) in the Xichang City (XC05). This phenomenon is consistent with the previous studies. For example, Sun et al. found that 6.25% of canine samples tested positive for CBuV, all of which were co-infected with CPV [44]. Similarly, Yao et al. found that 37.5% (3/8) of CBoV-positive samples were co-infected with CPV [7]. In this study, most virus-positive dogs were unvaccinated or incompletely vaccinated, including 91.67% of CPV-2–positive dogs, 75.0% of CBoV-positive dogs, and all CBuV-positive dogs. Similarly, a recent epidemiological study in China reported that incomplete or insufficient vaccination coverage was closely associated with the high prevalence and sustained circulation of canine parvovirus [50]. Therefore, these observations further underscore the importance of timely and complete vaccination in preventing canine viral enteric infections.

The distribution of CPV genotypes shows significant geographical variation, which highlights their dynamic evolutionary characteristics. Our genetic analysis confirmed that the CPV-2c genotype is dominant in Sichuan, accounting for 95.8% (46/48) of CPV-positive strains. Previous studies in Sichuan Province have indicated that CPV-2a and CPV-new 2a were the predominant circulating strains between 2011 and 2020 [22,25,28,34]. However, our data from 2020 to 2022 clearly highlighted that CPV-2c was the most prevalent genotype. This finding is consistent with the national trend observed in the provinces of Henan and Jilin [19,21], further establishing CPV-2c as the dominant variant in China. Additionally, we further analyzed the amino acid variations in the VP2 protein of CPV-2 strains. All Sichuan strains in our study belonged to the 'Asian CPV-2c', characterized by VP2 substitutions (Phe267Tyr, Tyr324Ile, and Gln370Arg), which distinguish them from Western CPV-2c strains in terms of their phylogenetic classification [51]. Phylogenetic analysis demonstrated that the Asian clade is subdivided into Asia-1.1 and Asia-1.2, primarily due to the I447M substitution. Within the Asia-1.1 subclade, two strains (MY21 and CD20) carried the Ala5Gly mutation and showed close phylogenetic relationships with Sichuan isolates from 2016 and 2017, indicating the sustained circulation of CPV-2c with Ala5Gly in Sichuan. The Ala5Gly mutation was first reported in Beijing, China, in 2015 [52]. Statistical analysis of Chinese Ala5Gly mutant strains revealed that this mutation first appeared in isolates from 2013 and has increased in prevalence in recent years, with occasional detections in Italy and Nigeria [21]. Except for Ala5Gly, other key mutations also were observed in our study, such as Ile447Met and Thr440Ala. In the Asian 1.2, both our LS03 strain and the new Vietnamese CPV-2c strain carried the Ile447Met mutation. However, the Vietnam new CPV-2c strain also carried Ala5Gly, whereas LS03 did not. This may explain why they are in the same major branch, but on distinct sub-branches. A previous study found that Vietnamese

CPV-2c strains had a unique non-synonymous Ile447Met mutation in VP2 genes, which was subsequently detected in Henan Province, China [21]. This suggests that the strain carried Ile447Met mutation may have been transmitted to China through the pet trade and population migration. Furthermore, we identified the Thr440Ala mutation in CPV-2c strains (CD04). At present, Thr440Ala mutation is only found in CPV-2a/2b and has been reported worldwide, including in Africa, India and China [53–56]. However, the detection of the Thr440Ala mutation in CPV-2c in our study suggests a possible genetic recombination event which deserves further investigation to understand its origin and implications.

Our study revealed the co-circulation of both CBoV-1 and CBoV-2 in Sichuan province. This finding contrasts with reports from Guangdong and Heilongjiang, where only CBoV-2 was detected [7,31], suggesting a geographical variation in the genetic diversity of CBoV across China. Regarding CBuV, our phylogenetic analysis identified two clusters, with the two Ya'an strains exhibiting significant genetic differences. Within cluster 2, our YA11 strain clustered with the Italian strain MT5402. Interestingly, the result is similar with the phylogenetic pattern observed in a study from Thailand, where two Thailand CBuV strains also formed a clade (subgroup B) with the same Italian MT5402 strain [57]. This specific genetic link between the Asian strains (from our study and from Thailand) and the Italian MT5402 strain suggests that they may have the same common ancestor.

This study has several limitations: (1) The modest sample size may limit the statistical power of our phylogenetic analysis. (2) The uneven sampling across regions (e.g., only 15 samples from Xichang) may not accurately reflect the prevalence and diversity of viruses across the entire study area. (3) Only diarrheic dogs' samples (not including health dogs' samples) were collected, which may introduce selection bias. Therefore, the detection rates reported here should not be interpreted as population-level prevalence but rather as the occurrence of viral infections among clinically affected dogs. Future studies including healthy control populations and population-based sampling would be necessary to assess statistical associations with diarrhea.

5. Conclusions

In this study, we have carried out a molecular characterization of CPV, CBoV and CBuV in dogs with enteric disease in Sichuan, China, with CPV-2c being the predominant pathogen (46 strains). Molecular and phylogenetic analyses confirmed that all CPV-2c strains belonged to the 'Asian CPV-2c' lineage, which has become the dominant genotype in Sichuan. Furthermore, the co-circulation of CBoV-1 and CBoV-2 reveals a greater genetic diversity of bocaviruses in Sichuan. This study is the first to report the identification of CBoV and CBuV in dogs in Sichuan province. Our findings highlight the increasing detection rates and genetic diversity of canine parvoviruses. Therefore, continuous monitoring the incidence and molecular characteristics of CPV-2c and these novel parvoviruses (CBoV and CBuV) is fundamental to controlling their spread.

Supplementary Materials: The following supporting information can be downloaded at: https://www.mdpi.com/article/10.3390/vetsci13010041/s1, Table S1: PCR protocols for the detection of CPV, CBoV, and CBuV in this study; Figure S1: original PCR images; Table S2: Amino acid variations in the VP2 protein of CPV-2 strains in this study and reference strain in this study; Table S3: Nucleotide identity among the CPV-2c strains in this study; Table S4: Amino acid identity among the CPV-2c strains in this study.

Author Contributions: Conceptualization, Z.Z. (Zhijun Zhong); methodology, Z.Z. (Zhijun Zhong); software, Z.Z. (Zhijun Zhong), R.H., X.L. (Xiaoqi Li), Y.Z., G.P., H.L. and Z.Z. (Ziyao Zhou); validation, X.L. (Xin Lei), M.W., X.L. (Xiaoqi Li), G.P., H.L. and Z.Z. (Ziyao Zhou); investigation, S.L., S.S., Y.H., I.L. and Y.Z.; data curation, S.L., S.S., Y.Z. and X.L. (Xiaoqi Li); writing—original draft preparation, S.L.,Y.Z.,

Y.H. and S.S.; writing—review and editing, X.L. (Xin Lei) and Z.Z. (Zhijun Zhong);visualization, S.L., M.C. and K.Z. All authors have read and agreed to the published version of the manuscript.

Funding: This work was supported by the National Key Research and Development Program of China (2024YFD1800202), and the Study on Key Technologies for Conservation of Wild Giant Panda Populations and Its Habitats within Giant Panda National Park System (CGF2024001).

Institutional Review Board Statement: This work was approved by the Sichuan Agricultural University Animal Ethical and Welfare Committee (ETHICS number: 20200583, October 9, 2020). All procedures involving animals were conducted in accordance with relevant animal welfare guidelines and regulations in China.

Informed Consent Statement: Informed consent was obtained from all subjects involved in the study.

Data Availability Statement: The original contributions presented in this study are included in the article and supplementary material. The gene sequences obtained in this study are openly available in the INSDC database under the following accession numbers: CPV (MZ857180-MZ857186, OP779632-OP779673, OQ030256-OQ030269), CBoV (OQ030256-OQ030263) and CBuV: (OQ030264-OQ030269, OQ020369). Further inquiries can be directed to the corresponding authors.

Conflicts of Interest: The authors declare no conflicts of interest.

References

1. Wang, H.; Li, Z.; Li, C.; Ma, Y.; Sun, Q.; Zhang, H.; Niu, G.; Wei, J.; Yao, H.; Ma, Z. Viral Metagenomic Analysis of the Fecal Samples in Domestic Dogs (*Canis lupus familiaris*). *Viruses* **2023**, *15*, 685. [CrossRef]
2. Kuzi, S.; Zgairy, S.; Byrne, B.A.; Suchodolski, J.; Turjeman, S.C.; Park, S.Y.; Aroch, I.; Hong, M.; Koren, O.; Lavy, E. Giardiasis and diarrhea in dogs: Does the microbiome matter? *J. Vet. Intern. Med.* **2024**, *38*, 152–160. [CrossRef]
3. Liu, S.; Zhou, B.; Liu, L.; Zhong, J.; Zhang, X.; Jiang, W.; Liu, H.; Zhou, Z.; Peng, G.; Zhong, Y.; et al. Effects and Microbiota Changes Following Oral Lyophilized Fecal Microbiota Transplantation Capsules in Canine with Chronic Enteropathy After Parvovirus Infection: Case Report. *Vet. Sci.* **2025**, *12*, 909. [CrossRef] [PubMed]
4. Qi, S.S.; Zhao, J.J.; Guo, D.H.; Sun, D. A Mini-Review on the Epidemiology of Canine Parvovirus in China. *Front. Vet. Sci.* **2020**, *7*, 5. [CrossRef]
5. Bodewes, R.; Lapp, S.; Hahn, K.; Habierski, A.; Förster, C.; König, M.; Wohlsein, P.; Osterhaus, A.D.; Baumgärtner, W. Novel canine bocavirus strain associated with severe enteritis in a dog litter. *Vet. Microbiol.* **2014**, *174*, 1–8. [CrossRef] [PubMed]
6. Di Martino, B.; Sarchese, V.; Di Profio, F.; Palombieri, A.; Melegari, I.; Fruci, P.; Aste, G.; Bányai, K.; Fulvio, M.; Martella, V. Genetic heterogeneity of canine bufaviruses. *Transbound. Emerg. Dis.* **2021**, *68*, 802–812. [CrossRef]
7. Decaro, N.; Martella, V.; Elia, G.; Desario, C.; Campolo, M.; Lorusso, E.; Cavalli, A.; Buonavoglia, C. Enteric viruses associated with diarrhoea in dogs. *Vet. Microbiol.* **2005**, *108*, 19–27.
8. Wang, J.Y.; Yan, Z.R.; Liu, H.R.; Wang, W.J.; Liu, Y.K.; Zhu, X.; Tian, L.; Zhao, J.J.; Peng, Q.S.; Bi, Z.W. Prevalence and Molecular Evolution of Parvovirus in Cats in Eastern Shandong, China, between 2021 and 2022 Transboundary and Emerging. *Diseases* **2024**, *2024*, 5514806.
9. Balboni, A.; Niculae, M.; Di Vito, S.; Urbani, L.; Terrusi, A.; Muresan, C.; Battilani, M. The detection of canine parvovirus type 2c of Asian origin in dogs in Romania evidenced its progressive worldwide diffusion. *BMC Vet. Res.* **2021**, *17*, 206. [CrossRef]
10. Carrino, M.; Tassoni, L.; Campalto, M.; Cavicchio, L.; Mion, M.; Corrò, M.; Beato, M.S. Molecular Investigation of Recent Canine Parvovirus-2 (CPV-2) in Italy Revealed Distinct Clustering. *Viruses* **2022**, *14*, 917. [CrossRef]
11. Mo, R.; Huang, P.; Zeng, Y.; Sun, Z.; Wang, S.M.; Wang, Z.S.; Pan, Z.H.; Wang, K.; Zhao, Y. Genetic evolution and genotype analysis of VP2 gene of canine parvovirus in Changchun. *Chin. J. Etiol.* **2020**, *15*, 302–305. [CrossRef]
12. Zhou, H.Z.; Su, X.; Lin, L.L.; Qi, Q.; Zhang, J.; Xu, F.Z.; Yang, B. Sequence analysis of VP2 gene of canine parvovirus and isolation and identification of BJ-1 strain. *Adv. Vet. Med.* **2020**, *41*, 21–26. [CrossRef]
13. Zhang, R.Z.; Yang, S.T.; Zhang, W.; Zhang, T.; Xie, Z.J.; Feng, H.; Wang, S.J.; Xia, X.Z. Phylogenetic analysis of the VP2 gene of canine parvoviruses circulating in China. *Virus Genes* **2010**, *40*, 397–402. [CrossRef]
14. Ohshima, T.; Hisaka, M.; Kawakami, K.; Kishi, M.; Tohya, Y.; Mochizuki, M. Chronological analysis of canine parvovirus type 2 isolates in Japan. *J. Vet. Med. Sci.* **2008**, *70*, 769–775. [CrossRef]
15. Liu, C.; Gao, J.; Li, H.; Sun, F.; Liang, H.; Liu, H.; Yi, J. Phylogenetic Characteristics of Canine Parvovirus Type 2c Variant Endemic in Shanghai, China. *Viruses* **2021**, *13*, 2257. [CrossRef] [PubMed]
16. Li, J.; Cheng, B.; Li, Z.; Cui, Y.; Yang, H.; Liu, W.; Zhang, C.; Yu, Y. Detection and molecular epidemiology of canine parvovirus and identification of highly pathogenic CPV-2c isolates from Shandong, China. *Virus Genes* **2025**, *61*, 7–109. [CrossRef]

https://doi.org/10.3390/vetsci13010041

17. Wang, Z.; Jiang, Y.; Xin, T.; Yuan, W.; Guo, X.; Zhou, H.; Zhu, H.; Jia, H. Genetic characterization and evolutionary analysis of canine parvovirus in Tangshan, China. *Arch. Virol.* **2022**, *167*, 2263–2269. [CrossRef]
18. Miao, B.; Chen, S.; Zhang, X.; Ma, P.; Ma, M.; Chen, C.; Zhang, X.; Chang, L.; Du, Q.; Huang, Y.; et al. T598 and T601 phosphorylation sites of canine parvovirus NS1 are crucial for viral replication and pathogenicity. *Vet. Microbiol.* **2022**, *264*, 109301. [CrossRef]
19. Jiang, H.; Yu, Y.; Yang, R.; Zhang, S.; Wang, D.; Jiang, Y.; Yang, W.; Huang, H.; Shi, C.; Ye, L.; et al. Detection and molecular epidemiology of canine parvovirus type 2 (CPV-2) circulating in Jilin Province, Northeast China. *Comp. Immunol. Microbiol. Infect. Dis.* **2021**, *74*, 101602. [CrossRef]
20. Sun, F.; Huang, X.; Li, B.; Zhou, H.; Long, J.; Qin, Y.; Ouyang, K.; Wei, Z.; Huang, W.; Chen, Y. Prevalence and molecular characterization of parvoviruses in dogs and domestic cats in Guangxi, China: New emergence of CPV-2c in cats. *Top. Companion Anim. Med.* **2025**, *68*, 101008. [CrossRef] [PubMed]
21. Fu, P.; He, D.; Cheng, X.; Niu, X.; Wang, C.; Fu, Y.; Li, K.; Zhu, H.; Lu, W.; Wang, J.; et al. Prevalence and Characteristics of Canine Parvovirus Type 2 in Henan Province, China. *Microbiol. Spectr.* **2022**, *10*, e0185622. [CrossRef]
22. Xu, J.; Guo, H.C.; Wei, Y.Q.; Shu, L.; Wang, J.; Li, J.S.; Cao, S.Z.; Sun, S.Q. Phylogenetic analysis of canine parvovirus isolates from Sichuan and Gansu provinces of China in 2011. *Transbound. Emerg. Dis.* **2015**, *62*, 91–95. [CrossRef]
23. Geng, Y.F.; Guo, D.H.; Li, C.Q.; Wang, E.Y.; Wei, S.; Wang, Z.H.; Yao, S.; Zhao, X.W.; Su, M.J.; Wang, X.Y.; et al. Co-Circulation of the Rare CPV-2c with Unique Gln370Arg Substitution, New CPV-2b with Unique Thr440Ala Substitution, and New CPV-2a with High Prevalence and Variation in Heilongjiang Province, Northeast China. *PLoS ONE* **2015**, *10*, e0137288. [CrossRef] [PubMed]
24. Wu, H.; Li, X.; Wang, L.; Liu, Y.; Tian, K. Molecular epidemiological survey of canine parvovirus in domestic dogs in four provinces, China. *Virusdisease* **2018**, *29*, 113–117. [CrossRef] [PubMed]
25. Cao, X.F.; Peng, G.N.; Huang, X.M.; Wang, W.B.; Tian, Y.N.; Xu, X.Y.; Lei, S.S.; Yang, Y.Y.; Gong, C.; Yang, K.X.; et al. Prevalence of parvovirus and phylogenetic analysis of VP2 gene in dogs with diarrhea. *Acta Virol.* **2016**, *32*, 733–739. [CrossRef]
26. Liu, X.; Ren, Z.; Zhou, Z.; Shen, L.; Geng, Y.; Zhao, J.; Zhong, Z. Molecular characterization of canine parvovirus isolates circulating in Southwest China. *Virus Genes.* **2013**, *47*, 248–255.
27. Zhang, Q.Y.; Qiu, Y.; Pan, Z.H.; Wang, S.C.; Wang, B.; Wu, W.K.; Yu, J.M.; Yi, Y.; Sun, F.L.; Wang, K.C. Genome sequence characterization of canine parvoviruses prevalent in the Sichuan province of China. *Transbound. Emerg. Dis.* **2019**, *66*, 897–907. [CrossRef]
28. Zhong, Z.; Liang, L.; Zhao, J.; Xu, X.; Cao, X.; Liu, X.; Zhou, Z.; Ren, Z.; Shen, L.; Geng, Y.; et al. First isolation of new canine parvovirus 2a from Tibetan mastiff and global analysis of the full-length VP2 gene of canine parvoviruses 2 in China. *Int. J. Mol. Sci.* **2014**, *15*, 12166–12187. [CrossRef]
29. Sobhy, N.M.; Armién, A.G.; Pilot, J.; Singh, V.; Padhi, S.K.; Muldoon, D.; Goyal, S.M.; Mor, S.K. Detection and molecular characterization of astro and bocaviruses in dogs in Minnesota. *Comp. Immunol. Microbiol. Infect. Dis.* **2021**, *81*, 101740. [CrossRef]
30. Acar, G.; Bostan Yoru, H.; Polat, E.; Aslan, A.N.; Aydin, H.; Timurkan, M.O. Isolation and molecular identification of canine bufavirus: A novel enteric pathogen of dogs. *Arch. Virol.* **2025**, *170*, 146. [CrossRef]
31. Zhai, S.L.; Lin, T.; Chen, Y.W.; Liu, R.; Lv, D.H.; Wen, X.H.; Zhu, X.L.; Wang, D.; Li, F.; Wei, W.K. First complete genome sequence of canine bocavirus 2 in mainland China. *New Microbes New Infect.* **2017**, *18*, 47–49. [CrossRef]
32. Zhang, W.; Li, L.; Deng, X.; Kapusinszky, B.; Pesavento, P.A.; Delwart, E. Faecal virome of dogs with diarrhoea. *Vet. Microbiol.* **2011**, *148*, 1–7.
33. Choi, J.W.; Lee, K.H.; Lee, J.I.; Lee, M.H.; Lee, K.K.; Oem, J.K. Genetic characteristics of canine bocaviruses in Korean dogs. *Vet. Microbiol.* **2015**, *179*, 177–183. [CrossRef]
34. Kapoor, A.; Mehta, N.; Dubovi, E.J.; Simmonds, P.; Govindasamy, L.; Medina, J.L.; Lipkin, W.I. Characterization of novel canine bocaviruses and their association with respiratory disease. *J. Gen. Virol.* **2012**, *93*, 341–346. [CrossRef]
35. Piewbang, C.; Jo, W.K.; Puff, C.; Ludlow, M.; van der Vries, E.; Banlunara, W.; Rungsipipat, A.; Kruppa, J.; Jung, K.; Techangamsuwan, S.; et al. Canine Bocavirus Type 2 Infection Associated with Intestinal Lesions. *Vet. Pathol.* **2018**, *55*, 434–441. [CrossRef] [PubMed]
36. Guo, D.; Wang, Z.; Yao, S.; Li, C.; Geng, Y.; Wang, E.; Sun, D. Epidemiological investigation reveals genetic diversity and high co-infection rate of canine bocavirus strains circulating in Heilongjiang province, Northeast China. *Res. Vet. Sci.* **2016**, *106*, 7–13. [CrossRef]
37. Lau, S.K.P.; Woo, P.C.Y.; Yeung, H.C.; Teng, J.L.L.; Wu, Y.; Bai, R.; Fan, R.Y.Y.; Chan, K.H.; Yuen, K.Y. Identification and characterization of bocaviruses in cats and dogs reveals a novel feline bocavirus and a novel genetic group of canine bocavirus. *J. Gen. Virol.* **2012**, *93*, 1573–1582. [CrossRef]
38. Martella, V.; Lanave, G.; Mihalov-Kovács, E.; Marton, S.; Varga-Kugler, R.; Kaszab, E.; Di Martino, B.; Camero, M.; Decaro, N.; Buonavoglia, C.; et al. Novel Parvovirus Related to Primate Bufaviruses in Dogs. *Emerg. Infect. Dis.* **2018**, *24*, 1061–1068. [CrossRef] [PubMed]
39. Sun, W.; Zhang, S.; Huang, H.; Wang, W.; Cao, L.; Zheng, M.; Yin, Y.; Zhang, H.; Lu, H.; Jin, N. First identification of a novel parvovirus distantly related to human bufavirus from diarrheal dogs in China. *Virus Res.* **2019**, *265*, 127–131. [CrossRef]
40. Li, J.; Cui, L.; Deng, X.; Yu, X.; Zhang, Z.; Yang, Z.; Delwart, E.; Zhang, W.; Hua, X. Canine bufavirus in faeces and plasma of dogs with diarrhoea, China. *Emerg. Microbes Infect.* **2019**, *8*, 245–247. [CrossRef] [PubMed]
41. Ganji, V.K.; Buddala, B.; Yella, N.R.; Putty, K. First report of canine bufavirus in India. *Arch. Virol.* **2022**, *167*, 1145–1149. [CrossRef] [PubMed]

https://doi.org/10.3390/vetsci13010041

42. Abayli, H.; Aslan, O.; Tumer, K.C.; Can-Sahna, K.; Tonbak, S. Investigation of canine chaphamaparvovirus, canine bufavirus, and canine adenovirus in dogs with diarrhea: First report of novel canine bufavirus in Turkey. *Virus Genes* **2023**, *59*, 427–436. [CrossRef]
43. Shao, R.; Zheng, F.; Cai, S.; Ji, J.; Ren, Z.; Zhao, J.; Wu, L.; Ou, J.; Lu, G.; Li, S. Genomic sequencing and characterization of a novel group of canine bufaviruses from Henan province, China. *Arch. Virol.* **2020**, *165*, 2699–2702. [CrossRef] [PubMed]
44. Wang, Y.; Guo, X.; Zhang, D.; Sun, J.; Li, W.; Fu, Z.; Liu, G.; Li, Y.; Jiang, S. Genetic and phylogenetic analysis of canine bufavirus from Anhui Province, Eastern China. *Infect. Genet. Evol.* **2020**, *86*, 104600. [CrossRef]
45. Ferrara, G.; Chianese, M.; Pagnini, U.; Iovane, G.; Montagnaro, S. Detection of old and recently discovered parvoviruses in dogs with enteritis. *Res. Vet. Sci.* **2021**, *137*, 18–23.
46. Ferrara, G.; Chianese, M.; Pagnini, U.; Iovane, G.; Montagnaro, S. Detection of old and recently discovered parvoviruses in apparently healthy dogs in Campania region, Italy. *Res. Vet. Sci.* **2025**, *196*, 105887. [CrossRef] [PubMed]
47. 2025 China Pet Industry White Paper. Available online: http://api.petdata.cn/index.html (accessed on 10 January 2025).
48. Diakoudi, G.; Lanave, G.; Capozza, P.; Di Profio, F.; Melegari, I.; Di Martino, B.; Pennisi, M.G.; Elia, G.; Cavalli, A.; Tempesta, M.; et al. Identification of a novel parvovirus in domestic cats. *Vet. Microbiol.* **2019**, *228*, 246–251. [CrossRef]
49. Chen, Y.; Wang, J.; Bi, Z.; Tan, Y.; Lv, L.; Zhao, H.; Xia, X.; Zhu, Y.; Wang, Y.; Qian, J. Molecular epidemiology and genetic evolution of canine parvovirus in East China, during 2018–2020. *Infect. Genet. Evol.* **2021**, *90*, 104780. [CrossRef]
50. Li, H.; Zhou, H.; Li, L.; Han, X.; Shang, Q.; Zhang, C.; Zhou, M.; Fu, Z.F.; Zhao, J.; Zhao, L. Epidemiological status and genome characteristics of canine parvovirus in China from 2022 to 2024. *Arch. Virol.* **2025**, *170*, 124. [CrossRef]
51. Chiang, S.Y.; Wu, H.Y.; Chiou, M.T.; Chang, M.C.; Lin, C.N. Identification of a novel canine parvovirus type 2c in Taiwan. *Virol. J.* **2016**, *13*, 160. [CrossRef]
52. Chen, B.; Zhang, X.; Zhu, J.; Liao, L.; Bao, E. Molecular Epidemiological Survey of Canine Parvovirus Circulating in China from 2014 to 2019. *Pathogens* **2021**, *10*, 588. [CrossRef]
53. Han, S.C.; Guo, H.C.; Sun, S.Q.; Shu, L.; Wei, Y.Q.; Sun, D.H.; Cao, S.Z.; Peng, G.N.; Liu, X.T. Full-length genomic characterizations of two canine parvoviruses prevalent in Northwest China. *Arch. Microbiol.* **2015**, *197*, 621–626. [CrossRef]
54. Mittal, M.; Chakravarti, S.; Mohapatra, J.K.; Chug, P.K.; Dubey, R.; Upmanuyu, V.; Narwal, P.S.; Kumar, A.; Churamani, C.P.; Kanwar, N.S. Molecular typing of canine parvovirus strains circulating from 2008 to 2012 in an organized kennel in India reveals the possibility of vaccination failure. *Infect. Genet. Evol.* **2014**, *23*, 1–6. [CrossRef]
55. Yi, L.; Tong, M.; Cheng, Y.; Song, W.; Cheng, S. Phylogenetic Analysis of Canine Parvovirus VP2 Gene in China. *Transbound. Emerg. Dis.* **2016**, *63*, e262–e269. [CrossRef] [PubMed]
56. Dogonyaro, B.B.; Bosman, A.M.; Sibeko, K.P.; Venter, E.H.; van Vuuren, M. Genetic analysis of the VP2-encoding gene of canine parvovirus strains from Africa. *Vet. Microbiol.* **2013**, *165*, 460–465. [CrossRef] [PubMed]
57. Charoenkul, K.; Thaw, Y.N.; Phyu, E.M.; Jairak, W.; Nasamran, C.; Chamsai, E.; Chaiyawong, S.; Amonsin, A. First detection and genetic characterization of canine bufavirus in domestic dogs, Thailand. *Sci. Rep.* **2024**, *14*, 4773. [CrossRef] [PubMed]

https://doi.org/10.3390/vetsci13010041

veterinary sciences

MDPI

Article

Phosphoproteomic Analysis Reveals Differences in Intercellular Spread Among Feline Herpesvirus Type 1 Mutants

Weiwei Lin [1,2], Xianglin Zhang [1], Qian Jiang [1], Hongtao Kang [1], Yijing Li [2,*], Honglin Jia [1] and Jiasen Liu [1,*]

1 State Key Laboratory of Animal Disease Control and Prevention, Harbin Veterinary Research Institute, Chinese Academy of Agricultural Sciences, Harbin 150069, China; linweiwei926@163.com (W.L.); 17873772373@163.com (X.Z.); jiangqian@caas.cn (Q.J.); kanghongtao@caas.cn (H.K.); jiahonglin@caas.cn (H.J.)

2 College of Veterinary Medicine, Northeast Agricultural University, Harbin 150030, China

* Correspondence: yijingli@163.com (Y.L.); neauljs@163.com (J.L.)

Simple Summary: This study aimed to elucidate how feline herpesvirus type 1 (FHV-1) serine/threonine protein kinase (pK) and thymidine kinase (TK) regulate viral replication through host protein phosphorylation. Using CRISPR/Cas9-generated pK^- and TK^- deletion mutants, we demonstrated that pK deletion resulted in significantly reduced plaque size and impaired replication capacity. Phosphoproteomic profiling revealed that pK deficiency specifically altered phosphorylation of spliceosome-associated proteins, whereas TK deletion predominantly affected the ATP-dependent chromatin remodeling pathway. Based on the experimental data, we hypothesize that PK and TK may regulate FHV-1 replication through distinct host signaling networks. Our work provides novel insights into FHV-1 pathogenesis and identifies potential targets for antiviral development.

Abstract: Feline herpesvirus-1 (FHV-1) is taxonomically classified within the family Herpesviridae, subfamily Alphaherpesvirinae, genus Varicellovirus, and species Felid alphaherpesvirus 1. The genome of FHV-1 is 135,797 bp in length and encodes 74 proteins. Among these proteins, serine/threonine protein kinase (pK) and thymidine kinase (TK) have been identified as potential virulence factors in alphaherpesviruses, although these kinases are dispensable for viral replication. As kinases, regulating phosphorylation modification is one of their functions, while the mechanism by which phosphorylation modification affects cell physiological functions and thereby influences viral replication remains unclear. In this study, we generated pK- and TK-deficient FHV-1 mutants by CRISPR/Cas9-mediated homologous recombination. The pK-deficient virus produced significantly smaller plaques than the TK-deficient virus. The replication kinetics of the pK-deficient virus were attenuated in multistep growth compared to the TK-deficient virus. These results indicate that deletion of the pK gene markedly reduces the replicative capacity of FHV-1. We applied data-independent acquisition (DIA) quantitative proteomics to profile changes in global protein expression and phosphorylation in F81 cells upon infection with TK^-, pK^-, and wild-type FHV-1 strain. The pK-deficient virus exhibited 3632 differentially phosphorylated proteins containing 11,936 modification sites; the TK-deficient virus showed 4529 differentially phosphorylated proteins with 19,225 phosphorylation sites. Functional characterization through Gene Ontology (GO) and Kyoto Encyclopedia of Genes and Genomes (KEGG) enrichment analyses identified significant involvement of phosphoproteins in spliceosome pathways in pK-deficient virus and ATP-dependent chromatin remodeling pathway in TK-deficient virus. Notably, several splicing regulators—including Ess-2 and CDK13, which modulate host spliceosomal function—displayed significantly reduced phosphorylation levels in pK-deficient viruses. A significant enrichment of ATP-dependent factors, such as SMARCA5 and RSF1, was observed in the TK-deficient virus. To our knowledge, this is the first investigation into the effects of FHV-1 infection on the host cell phosphoproteome. These data offer new insights into the phosphoregulatory circuits and signaling networks triggered by FHV-1 and may enhance our understanding of the FHV-1 replication mechanism.

Keywords: FHV-1; phosphoproteomics; differential phosphorylation; GO; KEGG

1. Introduction

Feline herpesvirus type 1 (FHV-1) is the primary causative agent of feline viral rhinotracheitis (FVR), a disease mainly characterized by upper respiratory and ocular signs [1]. FHV-1 belongs to the genus Varicellovirus, subfamily Alphaherpesvirinae, family Herpesviridae [2,3]. It is an enveloped virus with a linear double-stranded DNA genome organized into unique long (UL) and unique short (US) regions, flanked by internal and terminal repeats (IRS, TRS), forming the structure UL-IRS-US-TRS [4,5]. The complete genome spans 135,797 bp and encodes 74 viral proteins, including 11 glycoproteins [6].

Notably, the US3 gene encodes a serine/threonine protein kinase (pK) involved in virion assembly, cytoskeletal remodeling, and evasion of host innate immunity [7]. The pK gene is essential for efficient viral replication in vitro, as its deletion impairs the growth of multiple herpesviruses, including FHV-1 [8]. Deletion of US3 attenuates FHV-1 virulence, enhances interferon production, impairs establishment of latency in the trigeminal ganglion, and reduces subsequent viral shedding [9]. In addition to the US3 protein, thymidine kinase (TK) has been implicated in herpesvirus pathogenicity. While TK is not required for in vitro replication or immunogenicity in FHV-1, it is critical for replication and dissemination of pseudorabies virus in the central nervous system [10]. Partial or complete deletion of the TK gene has been shown to reduce virulence and neuroinvasiveness in several herpesviruses [11].

Protein phosphorylation represents a fundamental post-translational modification that orchestrates crucial biological processes, including signal transduction, molecular interactions, and enzymatic activity regulation [12,13]. This covalent modification significantly expands protein functional diversity, with approximately 30% of the proteome undergoing phosphorylation at one or more residues [14]. These modifications are catalyzed by protein kinases, which constitute nearly 2% of the eukaryotic genome [15]. In mammalian systems, phosphorylation primarily occurs on serine, threonine, and tyrosine residues. The reversible nature of this process—tightly regulated by kinases and phosphatases—enables dynamic control of cellular signaling networks. Substantial evidence demonstrates the indispensable role of phosphorylation in diverse cellular processes, including inflammasome assembly and activation [16], apoptosis [17], oncogenic proliferation [18,19], and virus–host interactions [20].

Previous studies have shown that in alphaherpesviruses, such as pseudorabies virus (PRV) and herpes simplex virus 1 (HSV-1), the pK protein directly phosphorylates core components of the m^{6}A mRNA methyltransferase complex, thereby suppressing its methylation activity [21]. In Marek's disease virus, the pK kinase controls replication and disease progression via direct phosphorylation of host HDAC1 and HDAC2 [22]. TK catalyzes the monophosphorylation of nucleoside analogues. Currently, the phosphorylation-related functions of the pK and TK genes remain incompletely characterized. Moreover, global phosphoproteomic changes induced by FHV-1 infection remain uncharacterized.

In this study, we generated targeted pK$^-$ and TK$^-$ mutants of FHV-1 using CRISPR/Cas9-mediated homologous recombination. Our assessment of viral growth curves and plaque sizes yielded results consistent with these earlier reports. These observations prompted us to further investigate the mechanisms underlying the differential replication efficiency among the mutant strains. Then, we performed phosphoproteomic sequencing and employed liquid chromatography-tandem mass spectrometry (LC-MS/MS) to profile host protein phosphorylation in response to infection with wild-type FHV-1 and its gene-deletion mutants. All experiments were conducted in three independent replicates to ensure reproducibility. Using the wild-type infection as a reference, we identified differentially phosphorylated proteins in mutant strains and performed functional annotation through Gene Ontology (GO), Kyoto Encyclopedia of Genes and Genomes (KEGG) pathway, and subcellular localization analyses. Our findings may elucidate the FHV-1 infection process and the mechanisms by which host phosphorylation dynamics block and limit viral replication.

2. Materials and Methods

2.1. Cells and Virus

Feline kidney cells (F81) were maintained in Dulbecco's Modified Eagle Medium (DMEM) supplemented with 10% (*v/vv/v*) fetal bovine serum (FBS), 100 U mL^{-1} penicillin, and 100 μg mL^{-1} streptomycin, and cultured at 37 °C in a humidified atmosphere containing 5% CO_2. Feline herpesvirus-1 strain QD-1 was isolated in this laboratory from the nasal swab of a kitten exhibiting clinically confirmed feline viral rhinotracheitis. For viral propagation, confluent F81 monolayers were inoculated with FHV-1 at a multiplicity of infection (MOI) of 0.01. When approximately 90% cytopathic effect (CPE) was observed, the culture supernatants were harvested and clarified for subsequent titration.

2.2. Selection of sgRNAs

Single-guide RNAs (sgRNAs) targeting the TK and pK sequences were designed using the online CRISPR design tool (https://www.zlab.bio/, Feng Zhang's Lab). The sgRNAs are listed in Table 1. The PX330 plasmid was digested using BbsI, and the CRISPR/Cas9-sgRNA plasmids were constructed. All of the constructs in this study were verified by sequencing.

Table 1. Sequences of the sgRNAs.

Gene	Primers	Sequence (5′-3′)
pK	sgRNA1-pK-F	caccgcgtggtatgctggacgatga
	sgRNA1-pK-R	aaactcatcgtccagcataccacg
	sgRNA2-pK-F	caccgcactctaaaggttcacccag
	sgRNA2-pK-R	aaacctgggtgaacctttagagtg
TK	sgRNA1-TK-F	caccgcatttacatagatggtgcct
	sgRNA1-TK-R	aaacaggcaccatctatgtaaatg
	sgRNA2-TK-F	caccgccttgggagcactgagacaa
	sgRNA2-TK-F	aaacttgtctcagtgctcccaagg

2.3. Construction of Recombinant Transfer Vectors

For deletion of the pK gene of FHV QD-1 by homologous recombination, a transfer vector was constructed by using two segments flanking the pK gene, pK-L and pK-R, as the recombination homologous arms. The left and right homology arms (pK-L and pK-R) were PCR-amplified from FHV QD-1 genomic DNA using the primer sets pKL-F/pKL-R and pKR-F/pKR-R, respectively. The EGFP expression cassette under the control of the CMV promoter and SV40 poly(A) signal was amplified from pEGFP-N1 with primers pK-EGFP-F and pK-EGFP-F using the primer set pK-EGFP-F and pK-EGFP-R. The pK-L and pK-R homology arms together with the EGFP expression cassette were directionally inserted between the BamHI and EcoRI sites of pUC19 to generate the final transfer vector pUC19-pKLR-EGFP. Following the identical cloning strategy, pUC19-TKLR-EGFP, a transfer plasmid designed for TK gene deletion, was constructed by inserting the corresponding TK-L and TK-R homology arms together with the EGFP expression cassette between the BamHI and EcoRI sites of pUC19. Primers used are listed in Table 2.

Table 2. Primers used to generate mutating fragments.

Primer	Sequence (5′–3′)	Used for
pKL-F	gaccatgattacgccaagcttgtctcgttgtcactctcgct	generation of pK^- mutating fragment
pKL-R	ttaccgtaagttatgtaacgggatccagtagagatcatcg	
pK-EGFP-F	cgttacataacttacggtaa	
pK-EGFP-R	gaacttggcatcgaaagtcataagatacattgatgagttt	
pKR-F	tgactttcgatgccaagttc	
pKR-R	aaacgacggccagtgaattcgttatgtatatcctcctccg	
TKL-F	accatgattacgccaagcttggttaacggacgatctgtga	generation of TK^- mutating fragment
TKL-R	ttaccgtaagttatgtaacgcgtctgatctgtgtatgatg	
TK-EGFP-F	cgttacataacttacggtaa	
TK-EGFP-F	atagggaacaccactaatgttaagatacattgatgagttt	
TKR-F	acattagtggtgttccctat	
TKR-R	aaacgacggccagtgaattcgcattccatcggccagtaat	

2.4. CRISPR/Cas9-Mediated Gene Deletion and Recombinant Virus Purification

F81 cells were transfected with 2 μg px330-pK-sgRNA and pUC19-pKLR-EGFP using PEI reagent according to the manufacturer's instructions. After 24 h, cells were inoculated with the FHV-1 QD-1 strain at an MOI of 0.01. At 48 h post-infection, cells and supernatant were collected and lysed by three consecutive freeze–thaw cycles to liberate intracellular virions. Recombinant FHV-1 was purified from the cell lysates by plaque purification in F81 cells overlaid with 1% low-melting-point agarose and 2% FBS. Following three rounds of plaque purification, EGFP-expressing plaques were readily visualized by fluorescence microscopy. The DNA of recombinant viruses was extracted using a TIANamp Genomic DNA Kit (TIANGEN, Beijing, China) and amplified using the specific primers pK-F1/pK-R1. The PCR product was sequenced, and the pK gene deletion and EGFP insertion were confirmed. The clonally purified, EGFP-expressing recombinant virus was named FHV QD-1 $pK^-/EGFP^+$. Similarly, the pUC19-TKLR-EGFP and px330-TK-sgRNA were transfected into F81cells, and the cells were infected with FHV-1 QD-1 strain 24 h after transfection. FHV QD-1 $TK^-/EGFP^+$ was screened by three rounds of purification and identification via PCR using primer pairs TK-F1/TK-R1 and sequencing. Primers used are listed in Table 3.

Table 3. Primers used to differentiate between wild-type and mutant viruses.

Primers	Sequence (5′–3′)	Used for
pK-F1	ccacacccatatcaacatc	Differentiation of pK^- mutant and wild type
pK-R	gtaacagagatccattgg	
TK-F1	aacctcacatgctaggtaca	Differentiation of TK^- mutant and wild type
TK-R1	ccagtcaacatcctcgatac	

2.5. Growth Kinetics and Plaque Assays of the Recombinant FHV-1

F81 cells were cultured in six-well plates and inoculated with FHV QD-1, $pK^-/EGFP^+$, and $TK^-/EGFP^+$ at 0.01 MOI, respectively. The supernatants and cells were collected at the indicated intervals. The virus titers were determined using the Reed–Muench method. Confluent F81 monolayers in six-well plates were inoculated with FHV QD-1, $pK^-/EGFP^+$, or TK^-/EGFP+ at 37 °C for 2 h, then overlaid with DMEM supplemented with 1% (*w*/*v*) low-melting-point agarose and 2% (*v/vv/v*) FBS, and incubated at 37 °C under 5% CO_2 for 72 h. The six-well culture plates were stained with 1% crystal violet in methanol for 1 h at room temperature before the plaques were observed.

2.6. Samples for LC-MS-MS Analysis

F81 cells were seeded in 75 cm^2 cell culture flasks and cultivated until achieving 80% confluency. The growth medium was removed, and cells were washed three times with sterile phosphate-buffered saline. Cells were infected with FHV QD-1, $pK^-/EGFP^+$, and $TK^-/EGFP^+$ at 0.01 MOI, incubated at 37 °C with 5% CO2, and harvested after 12 h for LC-MS-MS analysis. Uninfected cells were incubated in DMEM alone as a non-infected control (NC) group. To ensure adequate coverage of phosphosites, three biological replicates (a total of nine samples), including three FHV QD-1, three $pK^-/EGFP^+$, and three $TK^-/EGFP^+$, were used.

2.7. Protein Extraction

Cells were resuspended in lysis buffer (8M urea, 1% protease inhibitor, 1% phosphatase inhibitor, 3 μM Trichostatin A, and 50 mM Nicotinamide) and sonicated three times on ice using a high-intensity ultrasonicator, and precipitates were removed by centrifugation at 12,000× *g* for 10 min at 4 °C. The supernatant was collected following the manufacturer's protocol, and protein concentrations were quantified using a bicinchoninic acid (BCA) assay.

2.8. Protein Digestion and Phosphopeptide Enrichment

Following quantification, protein aliquots were subjected to reduction with 10 mM dithiothreitol (DTT) at 37 °C for 1 h and alkylation with 50 mM iodoacetamide (IAA) in the dark at room temperature for 1 h. Excess IAA was quenched by the addition of 5 mM DTT with incubation at room temperature for 10 min. The samples were then transferred to 30-kDa molecular weight cut-off (MWCO) filters and centrifuged. The filter units were subsequently washed twice with 100 μL of UA buffer (8 M urea in 100 mM Tris-HCl, pH 8.5) and three times with 100 μL of 50 mM ammonium bicarbonate (NH_4HCO_3) to remove residual denaturant. After buffer exchange, trypsin was added at a 1:50 (*w*/*w*) enzyme-to-protein ratio in 50 mM NH_4HCO_3, and digestion proceeded at 37 °C for 16–20 h. The resulting peptides were desalted using C18 cartridges, lyophilized, and reconstituted in 40 μL of 0.1% formic acid. Peptide concentration was determined by absorbance at 280 nm. For phosphopeptide enrichment, the HiSelect^{TM}TiO_2 Phosphopeptide Enrichment Kit (Thermo Fisher Scientific, Waltham, MA USA, A32993) was used according to the manufacturer's protocol. The enriched phosphopeptides were lyophilized and stored in 0.1% formic acid for subsequent LC-MS/MS analysis.

2.9. Nanoflow Ultra-High Performance Liquid Chromatography (UHPLC)

Chromatographic separation was performed using a nanoflow ultra-high performance liquid chromatography (UHPLC) system. The mobile phases consisted of 0.1% formic acid in water (phase A) and 0.1% formic acid in 80% acetonitrile (phase B). After equilibrating the column with 100% phase A, samples were loaded via an autosampler onto a trapping column and subsequently separated on an analytical column. The mass spectrometry gradient program is detailed in Table 4.

Table 4. Mass Spectrometry Gradient.

Time (min)	Flow (µL/min)	%B
0	0.4	6.7
0.7	0.4	6.7
1	0.4	7.2
11.3	0.4	24
17	0.4	36
18.5	0.4	55
19	0.4	99
24	0.4	99

2.10. Mass Spectrometric Analysis

Following separation by nanoflow liquid chromatography, samples were analyzed using an Orbitrap Astral mass spectrometer (Thermo Scientific) in positive ion mode. The total analysis time was 24 min. Detailed mass spectrometry parameters are listed in Table 5.

Table 5. Mass Spectrometry Parameter Settings Table.

Properties of Full MS	
General	
Runtime (min)	24
Polarity	Positive
Full MS	
Resolution	240,000
AGC target (%)	500
Maximum Injection Time (ms)	5
Scan Range (m/z)	380–980
Properties of DIA	
General	
Runtime (min)	24
Polarity	Positive
Default charge state	2
Advanced Peak Determination	Ture
MS2	
Resolution	80,000
Precursor Mass Range (m/z)	380–980
DIA Window Type	Auto
Isolation Window (m/z)	2
Number Of Scan Events	299
DIA Window Mode	m/z Range
Collision Energy Type	Normalized
HCD Collison Energy (%)	25
Detector Type	Astral
Scan Range (m/z)	150–2000
AGC target (%)	500
Maximum Injection Time (ms)	3
Microscans	1
Loop Control	Time
Time (s)	0.6

2.11. Mass Spectrometry Data Processing

A spectral library was constructed by searching against a protein sequence database in the FASTA format, which included the Felis catus (host) and FHV-1 (virus) proteomes. Subsequent analysis of data-independent acquisition (DIA) data was performed using

Spectronaut Pulsar software 20 (Biognosys AG, Schlieren, Switzerland). Following database searching against the feline-specific UniProt database, raw LC-MS/MS files were imported into Spectronaut for spectral library generation and comprehensive data processing. In order to test the reproducibility of experiments, phosphoprotein ratios in each replicate were quantified based on the summed intensity of the matched spectra. Pearson correlation coefficient and average relative standard deviation (RSD) were explored to evaluate the reproducibility of protein relative quantitation.

2.12. Data Filtering and Preprocessing

To ensure data quality, proteins and peptides were retained only if they were detected in more than 50% of samples within any experimental group. For the filtered dataset, missing values were imputed using the mean of respective sample groups. The resulting data then underwent median normalization followed by $\log_2$ transformation to generate a finalized set of high-confidence phosphoproteins and phosphopeptides.

2.13. Differential Expression Analysis

For the identification of significantly differentially phosphorylated peptides, we applied the following criteria: a phosphopeptide was considered significantly upregulated if it exhibited a fold change (FC) > 1.2 with a p-value < 0.05 (Student's t-test), and downregulated if it exhibited a fold change (FC) < 0.5 with a p-value < 0.05. The numbers of up- and downregulated phosphopeptides for each comparison were determined based on these thresholds.

2.14. Bioinformatic Analysis and Visualization

Phosphorylation motif enrichment was analyzed using the MOMO tool from the MEME Suite. Functional characterization of the phosphoproteome was performed through integrated bioinformatic approaches, including Gene Ontology (GO) term enrichment, Kyoto Encyclopedia of Genes and Genomes (KEGG) pathway analysis, and the OmicShare Tools platform. Data visualization was implemented using multiple specialized tools: density plots were generated with the ggplot2 package in R, protein correlation patterns were visualized using the corrplot package (v0.84) in R, and bar charts, along with pie charts, were created using GraphPad Prism (v8.0.2). The phosphorylated protein analysis was performed in three independent experiments. Data are presented as the mean $\pm$ standard deviation unless otherwise specified. Statistical significance was defined as $p < 0.05$ (GraphPad 6/ SPSS 20).

3. Results

3.1. Construction and Identification of the FHV-1 Mutant via CRISPR/Cas9-Mediated Homologous Recombination

Utilizing CRISPR/Cas9-mediated homologous recombination (Figure 1A), we generated a series of FHV-1 mutants. F81 cells were co-transfected with the pKL-CMV-EGFP-bGH polyA-pKR fragment and a sgRNA specific for the pK gene. At 24 h post-transfection, the cells were infected with the FHV-1 QD-1 strain at an MOI of 0.01. Following the emergence of green fluorescent plaques exhibiting characteristic herpesvirus cytopathic effects, the $pK^-/EGFP^+$ virus was isolated through three rounds of plaque purification (Figure 1B). The identity of the recombinant virus was then verified by PCR amplification and sequencing, which confirmed a distinct amplification profile compared to the wild-type strain. The size of the $pK^-/EGFP^+$ target gene is 1700 bp, while that of FHV-1 is 1200 bp. (Figure 1C). Finally, the sequence-confirmed recombinant virus, in which EGFP replaces the pK gene, was harvested and stored for subsequent experiments. The TK^- mutant was constructed following an identical procedure. The PCR identification results indicate the size of the pK⁻/EGFP⁺ target gene is 1700 bp, while that of FHV-1 is 1200 bp, which meets our expectations.

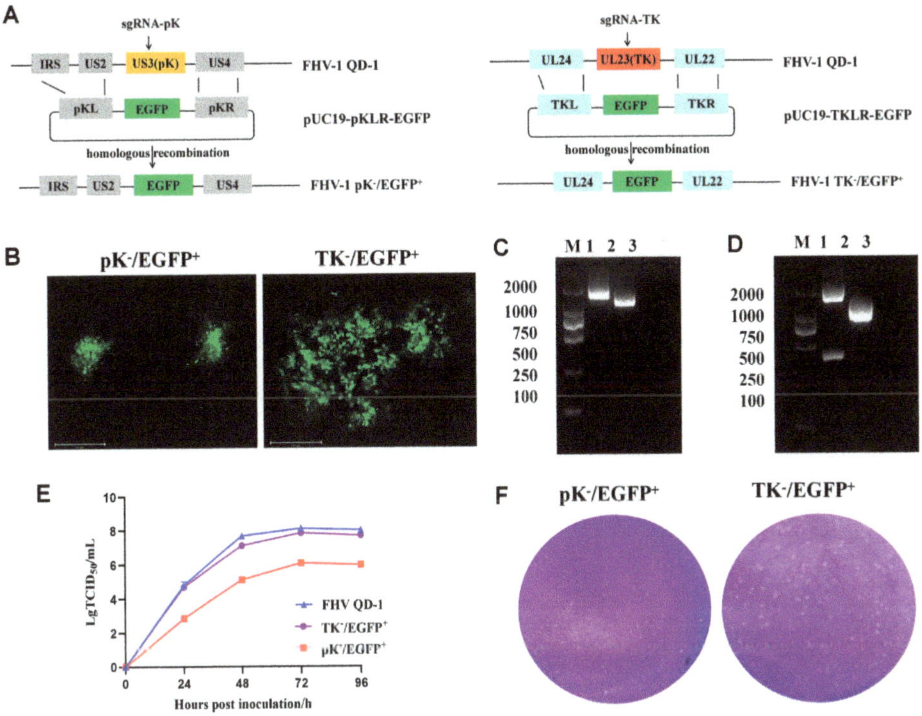

Figure 1. Construction and identification of recombinant FHV-1. (**A**) Schematic diagram of the recombinant FHV-1 development strategy. (**B**) Plaque purification of $pK^{-}/EGFP^{+}$ and $TK^{-}/EGFP^{+}$ in agarose-DMEM plates. (**C**) PCR confirmation of the pK gene deletion and the EGFP insertion. Lane M: DNA Molecular Weight Standard (DL2000); Lane 1: $pK^{-}/EGFP^{+}$; Lane 2: FHV-1 WT QD-1; Lane 3: blank control. (**D**) PCR confirmation of the TK gene deletion and the EGFP insertion. Lane M: DNA Molecular Weight Standard (DL2000); Lane 1: $TK^{-}/EGFP^{+}$; Lane 2: FHV-1 WT QD-1; Lane 3: blank control. (**E**) Multiple-step growth curves of FHV QD-1, $pK^{-}/EGFP^{+}$, and $TK^{-}/EGFP^{+}$ in F81 (0.01 MOI). (**F**) Plaque morphologies and sizes of F81 cells inoculated with the $pK^{-}/EGFP^{+}$, and $TK^{-}/EGFP^{+}$.: Original PCR/Gel Electrophoresis pictures see Figures S1 and S2.

To assess the impact of pK or TK deletion on FHV-1 biological properties, we examined viral replication kinetics and cell-to-cell spread. Using the FHV-1 QD-1 strain as a control, multistep growth curve analysis revealed that the $pK^{-}/EGFP^{+}$ mutant replicated more slowly than both the wild-type QD-1 and $TK^{-}/EGFP^{+}$ strains, requiring an extended time to achieve CPE in F81 cells, the peak titer of the $pK^{-}/EGFP^{+}$ mutant reached 10^6 $TCID_{50}/mL$, significantly lower than the $10^{8.3}$ $TCID_{50}/mL$ attained by the wild-type FHV-1 QD-1 strain (Figure 1E). The data confirm that pK gene deletion reduces FHV-1 replicative fitness. As shown in Figure 1B, the $pK^{-}/EGFP^{+}$ mutant formed significantly smaller plaques than $TK^{-}/EGFP^{+}$, indicating that pK deletion impairs intercellular viral spread.

3.2. Global Detection of Phosphosites in F81 Cells Infected with FHV-1 Mutants

Analysis of phosphorylation modifications revealed both single-site and multisite events. To quantify these modifications, a comparative analysis between the $pK^{-}/EGFP^{+}$

and WT FHV-1 groups was performed. As illustrated in Figure 2A, this analysis identified 22,099 phosphorylated peptides, 11,935 phosphorylation sites, and 3632 phosphorylated proteins. Regarding the distribution of phosphorylation types, serine residues accounted for the majority (77.1%) of all identified sites, followed by threonine (15.9%) and tyrosine (7.0%) (Figure 2B). Subsequently, subcellular localization prediction indicated that these phosphorylated proteins were primarily distributed in the nucleus and cytoplasm (Figure 2C). Furthermore, analysis of the phosphorylated peptides revealed that 11,677 contained a single phosphorylation site, while 254, 4, and 1 peptides contained two, three, and four sites, respectively (Figure 2D).

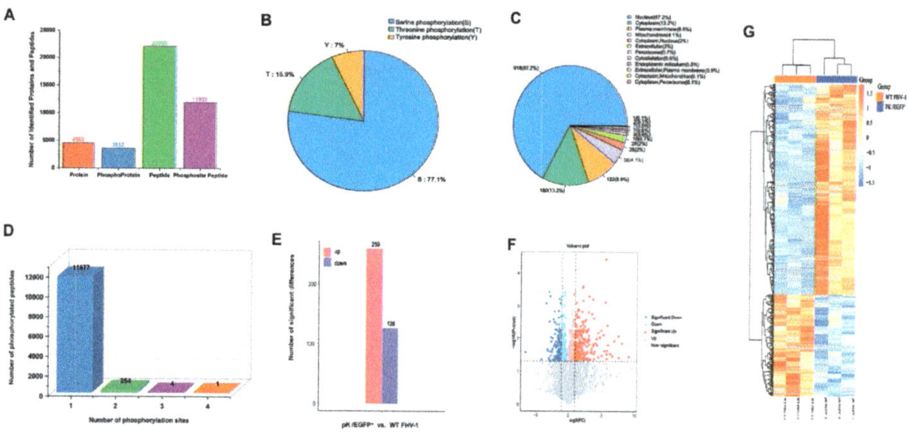

Figure 2. A quantitative overview of the phosphoproteomic analysis of phosphoproteins with differentially regulated phosphosites in pK$^-$/EGFP$^+$ and WT FHV-1 groups. (**A**) The number of phosphopeptides, phosphorylation sites, and phosphoproteins identified. (**B**) Quantitative statistics of the phosphorylation of serine, threonine, and tyrosine. (**C**) Prediction of the subcellular localization of phosphorylated proteins. (**D**) Statistical analysis of the number of modification sites on the phosphorylated peptide fragments. (**E**) Overview of differential phosphorylated phosphopeptides. (**F**) Volcano map analysis of the differential phosphorylated phosphopeptides. (**G**) Hierarchical clustering analysis of the differential phosphorylated phosphopeptides.

In a parallel comparative phosphoproteomic analysis of the TK$^-$/EGFP$^+$ and WT FHV-1 groups, a larger dataset was obtained, comprising 68,816 phosphopeptides, 19,225 phosphosites, and 4529 phosphorylated proteins (Figure 3A). Consistent with the trend observed above, phosphoserine (pS) residues were again predominant (77.9%), with phosphothreonine (pT) and phosphotyrosine (pY) constituting 17.8% and 4.3% of the sites, respectively (Figure 3B). Regarding cellular distribution, the phosphoproteins in this dataset were predicted to localize primarily to the nucleus, cytoplasm, and plasma membrane (Figure 3C). Finally, the distribution analysis of phosphosites showed that 17,360 peptides featured a single site, whereas 1582, 266, and 16 peptides contained two, three, and four sites, respectively (Figure 3D).

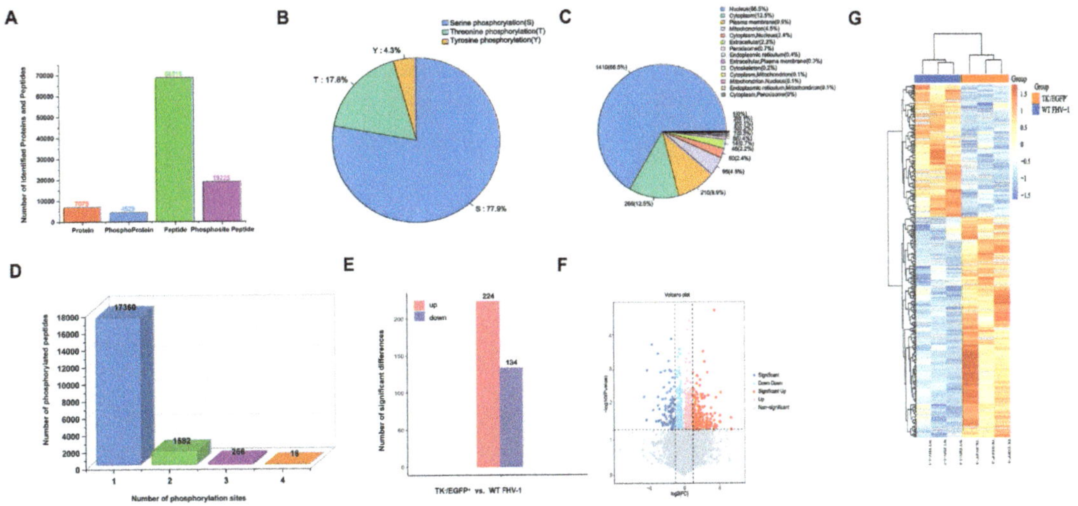

Figure 3. A quantitative overview of the phosphoproteomic analysis of phosphoproteins with differentially regulated phosphosites in TK^{-}/EGFP^{+} and WT FHV-1. (**A**) The number of phosphopeptides, phosphorylation sites, and phosphoproteins identified. (**B**) Quantitative statistics of the phosphorylation of serine, threonine, and tyrosine. (**C**) Prediction of the subcellular localization of phosphorylated proteins. (**D**) Statistical analysis of the number of modification sites on the phosphorylated peptide fragments. (**E**) Overview of differential phosphorylated phosphopeptides. (**F**) Volcano map analysis of the differential phosphorylated phosphopeptides. (**G**) Hierarchical clustering analysis of the differential phosphorylated phosphopeptides.

To further explore the differential phosphorylation of host proteins in response to infection by the wild-type strain compared to the gene deletion strains. Following pK^{-}/EGFP^{+} infection, 385 phosphorylated peptides were identified as significantly differentially expressed. Of these, 259 were significantly upregulated (fold change $\geq$ 1.5, $p \leq 0.05$), while 126 were significantly downregulated (ratio $\leq$ 0.67, $p \leq 0.05$) (Figure 2E). 224 upregulated and 134 downregulated differentially expressed proteins after TK^{-}/EGFP^{+} infection (Figure 3E). The overall distribution of these significant changes is displayed in volcano plots (Figures 2F and 3F). Subsequently, clustering analysis confirmed a clear separation between the phosphopeptides of the WT and gene deletion strain groups, underscoring the distinct impact of the viral genotype on the host phosphoproteome. Furthermore, among the differentially phosphorylated peptides, upregulated species were substantially more abundant than downregulated ones, a finding consistent with the trends observed in the volcano plots (Figures 2G and 3G).

3.3. Analysis of Phosphorylated Peptide Motifs

To investigate the sequence conservation surrounding the phosphorylation sites, we analyzed the flanking regions spanning six amino acids upstream to six amino acids downstream of each modified residue. This analysis revealed conserved sequence motifs and aided in predicting potential cognate kinases. Specifically, motif analysis of the 1276 phosphopeptides identified in both the pK^{-}/EGFP^{+} and WT FHV-1 groups showed that serine was the predominant phosphorylated residue (Figure 4A(1–3)(5–7)). Among these, six serine phosphorylation motifs were significantly enriched, with SP and RxxS being the most prevalent (Figure 4B). Furthermore, examination of all 2270 phosphopeptides identified 14 conserved motifs, including 9 phosphoserine (P-Ser) (Figure 5A(1–6)(8–9)) and 5 phosphothreonine (P-Thr) motifs (Figure 5A(7)(10–13)), in the TK^{-}/EGFP^{+} and WT

FHV-1 groups. These motifs exhibited distinct abundance patterns, with SP being the most widely distributed phosphoserine motif, followed by SxS and RxxS (Figure 5B).

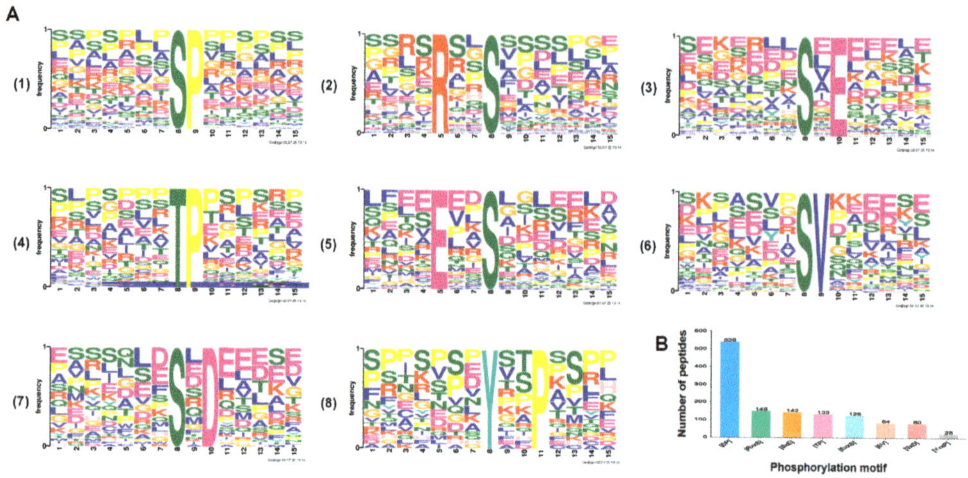

Figure 4. The phosphorylation motifs identified by the pK$^-$/EGFP$^+$ and WT FHV-1 groups. (**A**) Significantly enriched phosphorylation sequences. (**B**) The number of recognized peptides containing phosphorylated motifs.

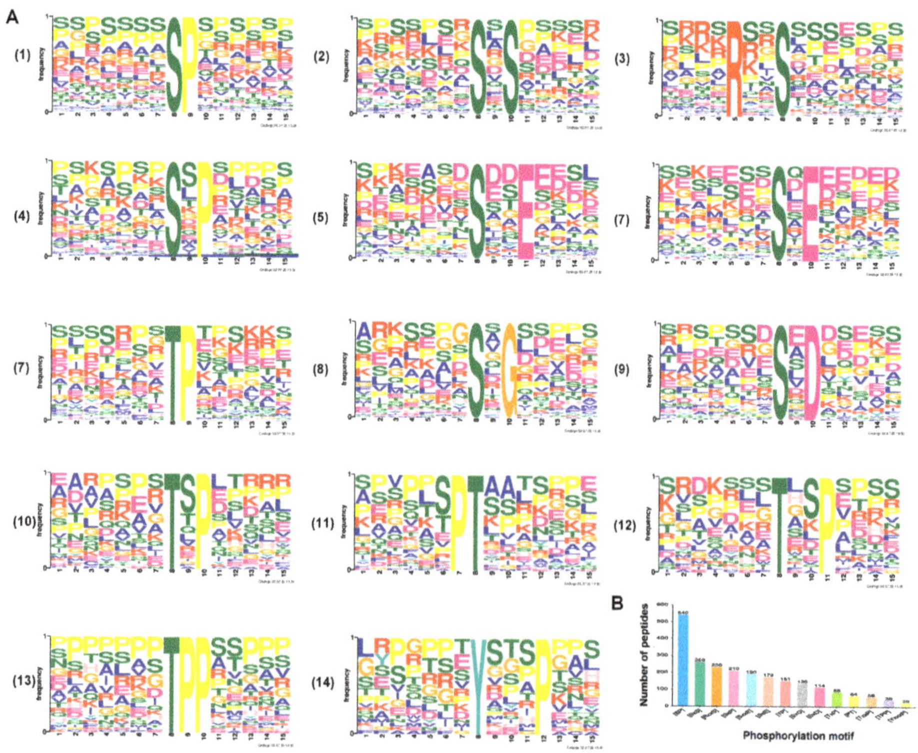

Figure 5. The phosphorylation motifs identified by the TK$^-$/EGFP$^+$ and WT FHV-1 groups. (**A**) Significantly enriched phosphorylation sequences. (**B**) The number of recognized peptides containing phosphorylated motifs.

3.4. Analysis of Phosphorylated Peptide Domain

In addition, domain prediction was performed on proteins harboring differentially expressed modified peptides. The phosphorylation levels of proteins associated with the RNA-binding domain superfamily and the RNA recognition motif (RRM) domain—both directly involved in RNA binding and processing—were observed to be elevated in the pK$^-$/EGFP$^+$ strain (Figure 6A). Upon phosphorylation, certain host RNA-binding proteins (RBPs) exhibit enhanced RNA-binding affinity or stability, enabling more efficient recognition and binding of viral RNA [23], followed by recruitment of degradation machinery for viral RNA clearance. Domain superfamilies implicated in chromatin remodeling and epigenetic regulation—including the SNF2 N-terminal, SNF2-like N-terminal, Bromodomain, and High Mobility Group box—were significantly upregulated in the TK$^-$/EGFP$^+$ strain (Figure 6B). Furthermore, subunits of many chromatin remodeling complexes, notably SWI/SNF, are regulated by phosphorylation. Phosphorylation upregulation directly enhances its ATPase activity or alters nucleosome positioning, thereby establishing an open chromatin environment at viral genome promoters to promote viral gene transcription [24,25]. Notably, the zinc finger PHD-type domain displayed distinct expression patterns between the mutants: it was upregulated in the pK$^-$/EGFP$^+$ strain, whereas in the TK$^-$/EGFP$^+$ strain, the extent of upregulation marginally surpassed that of downregulation. Given that the PHD zinc finger is a critical epigenetic reader of the histone code and plays essential roles in transcriptional regulation and DNA damage repair [26], its differential expression across viral strains suggests altered strategies in virus–host chromatin interactions during infection.

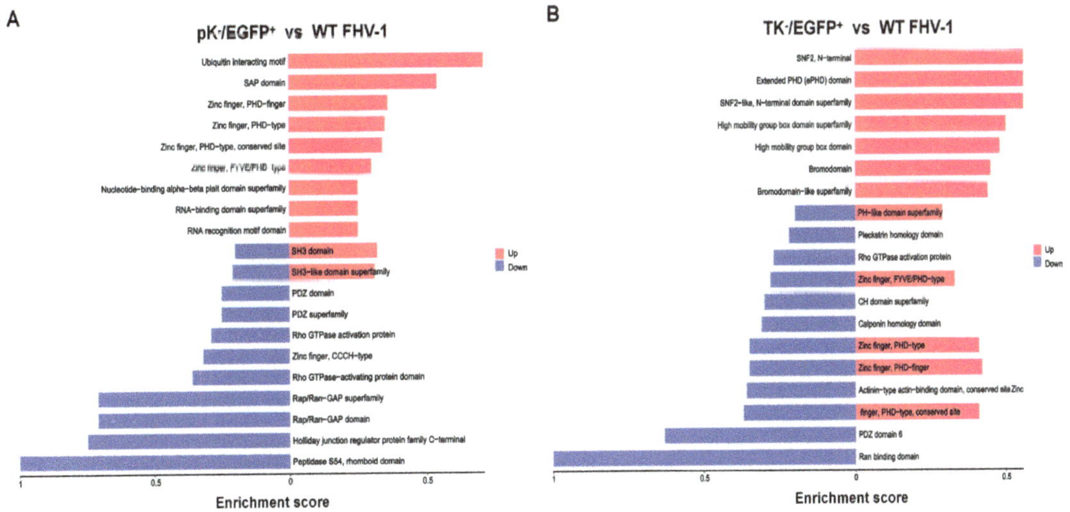

Figure 6. The structural domain prediction histograms. (**A**) The structural domain prediction histograms in the pK$^-$/EGFP$^+$ and WT FHV-1 groups. (**B**) The structural domain prediction histograms in the TK$^-$/EGFP$^+$ and WT FHV-1 groups.

3.5. Functional Enrichment of Proteins with Differentially Regulated Phosphosites

We performed Gene Ontology (GO) enrichment analysis on proteins associated with differential phosphorylation sites to systematically characterize their functional implications. The GO framework categorizes protein functions into three primary domains: Biological Process, Cellular Component, and Molecular Function. In the comparison between the pK$^-$/EGFP$^+$ and WT FHV-1 groups, proteins with downregulated phosphosites revealed a distinct functional profile. Specifically, within the Molecular Function category, the most significantly enriched terms included enzyme activator activity, transcription coregulator activity, GTPase activator activity, nucleoside-triphosphatase regulator activity, and GTPase regulator activity. Regarding cellular localization, the nuclear lumen and nucleoplasm were the most enriched Cellular Component terms. In the Biological Process domain, the top four enriched subcategories were regulation of cellular process, regulation of isotype switching, lens development in camera-type eye, and regulation of biological process (Figure 7A). Unlike the downregulated phosphosites, the enrichment results for proteins with upregulated phosphosites yielded distinct functional characteristics in both the Molecular Function and Cellular Component categories. Notably, DNA binding emerged as the most significantly enriched Molecular Function term, followed by nucleic acid binding. Within the Cellular Component category, the top five enriched terms were host cellular component, host intracellular region, host intracellular part, host cell part, and host intracellular membrane-bounded organelle (Figure 7B).

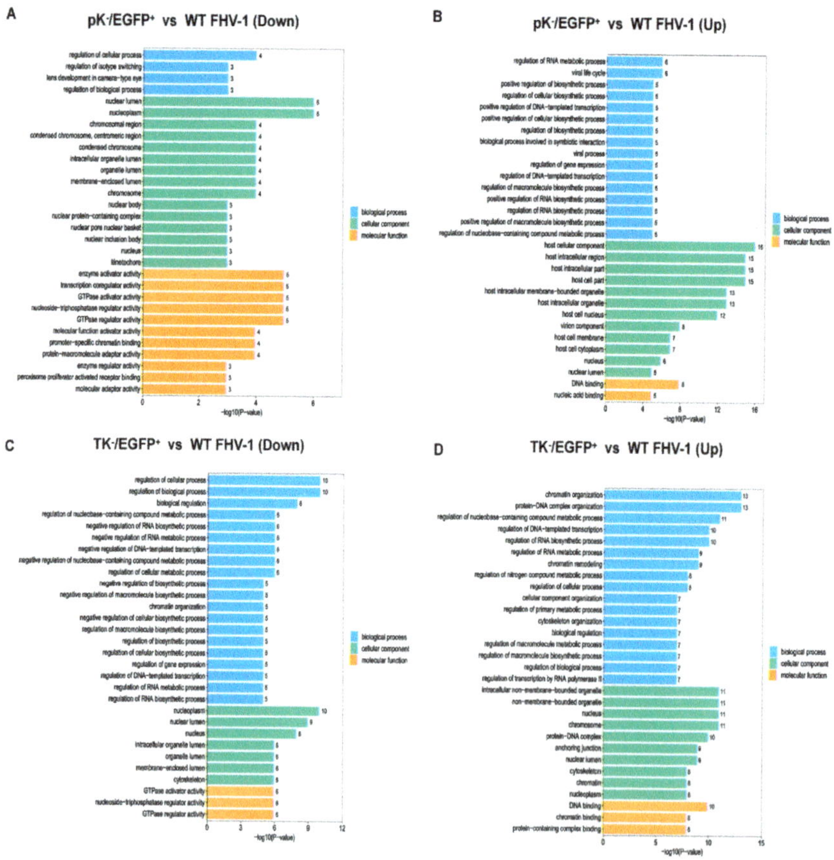

Figure 7. Gene Ontology functional enrichment analyses of proteins with differentially regulated phosphosites. (**A**) GO-based enrichment analysis of proteins with upregulated phosphosites in the $pK^{-}/EGFP^{+}$ and WT FHV-1 groups. (**B**) GO-based enrichment analysis of proteins with downregulated phosphosites in the $pK^{-}/EGFP^{+}$ and WT FHV-1 groups. (**C**) GO-based enrichment analysis of proteins with upregulated phosphosites in the $TK^{-}/EGFP^{+}$ and WT FHV-1 groups. (**D**) GO-based enrichment analysis of proteins with downregulated phosphosites in the $TK^{-}/EGFP^{+}$ and WT FHV-1 groups.

We next extended this analysis to the TK$^-$/EGFP$^+$ and WT FHV-1 comparison. For phosphoproteins with downregulated sites, GO enrichment again showed category-specific patterns. Nucleoplasm represented the most significantly enriched Cellular Component, with nuclear lumen ranking second. In parallel, regulation of cellular process and regulation of biological process were the primary terms enriched in the Biological Process category (Figure 7C). Conversely, proteins with upregulated phosphosites in this comparison displayed a contrasting functional signature. Specifically, the top three enriched Biological Process terms were chromatin organization, protein-DNA complex organization, and regulation of nucleobase-containing compound metabolic process. Meanwhile, the top four enriched Cellular Component categories were intracellular non-membrane-bounded organelle, non-membrane-bounded organelle, nucleus, and chromosome (Figure 7D).

To further explore the roles of differentially phosphorylated proteins and to uncover the molecular mechanisms underlying FHV-1 intercellular transmission, we used the Kyoto Encyclopedia of Genes and Genomes (KEGG) for pathway enrichment analysis. This approach identified distinct signaling and regulatory pathways across the mutant strains. A comparative analysis of viral mutants revealed distinct mechanisms of host manipulation. While the pK$^-$/EGFP$^+$ strain exhibited significant enrichment of the spliceosome pathway, implicating severe disruption of post-transcriptional regulation (Figure 8A), the TK$^-$/EGFP$^+$ mutant induced a major reprogramming of the host epigenetic landscape. This was evidenced by significant enrichment of the ATP-dependent chromatin remodeling pathway (Figure 8B), a feature markedly less pronounced in the pK$^-$/EGFP$^+$ strain.

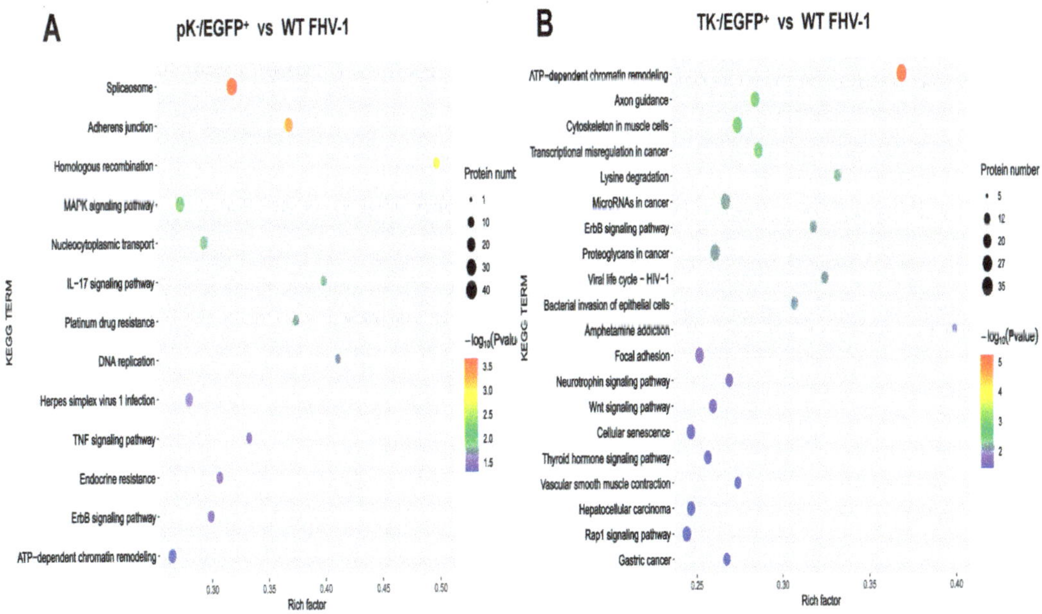

Figure 8. Kyoto Encyclopedia of Genes and Genomes (KEGG) pathway analysis of the differentially expressed proteins. (**A**) KEGG enrichment analysis of differentially expressed proteins in the comparison between the pK$^-$/EGFP$^+$ and WT FHV-1 groups (**B**) KEGG enrichment analysis of differentially expressed proteins in the comparison between the TK$^-$/EGFP$^+$ and WT FHV-1 groups.

4. Discussion

FHV-1 mutants with deletions in the pK or TK genes have been previously constructed and characterized in vitro [8,27]. In the present study, we also generated pK$^-$ and TK$^-$

deletion mutants and systematically characterized their biological properties. Our results recapitulated the established phenotype of the pK$^-$ and TK$^-$ mutant; the pK$^-$ mutant produced smaller plaques and exhibited slower replication kinetics than the TK$^-$ mutant. These are likely linked to its function in mediating nucleocapsid nuclear egress. Therefore, pK plays a critical role in promoting viral replication.

Viral replication emerges from complex host–pathogen interactions, with phosphorylation events orchestrated by protein kinases comprising a major regulatory axis in this process. The mechanistic basis by which the protein kinases pK and TK regulate phosphorylation during FHV-1 infection remains unclear. In this study, we systematically analyzed differentially regulated protein phosphorylation events following infection with distinct FHV-1 mutants to construct a comprehensive phosphoproteomic landscape. Our analysis identified 126 downregulated phosphorylation sites in the pK$^-$ mutant and 134 in the TK$^-$ mutant. Next, we performed a functional enrichment analysis on proteins with differentially regulated phosphorylation.

The pK$^-$ mutant profoundly reprograms the host cellular environment, primarily by disrupting central gene regulatory machinery and activating immune-inflammatory pathways. Among these, the most pronounced alteration was observed in the spliceosome pathway. Consistent with this, we observed a concomitant shift in the phosphorylation status of key regulators of host gene expression and inflammation, such as Ess-2, CDK13, and STAT5A.

Ess-2 is an RNA-binding protein primarily involved in pre-mRNA splicing, a crucial step in the regulation of gene expression [28]. Notably, recent studies have shown that many viral infections lead to the reprogramming of host cell splicing, which can be attributed to direct viral manipulation, virus-induced immune responses, or cellular damage [29–31]. A well-established example of this viral strategy is the HSV-1-encoded protein ICP27, which interacts with SAP145, a critical component of the U2 snRNP complex within the spliceosome [32]. This interaction promotes the nuclear export of viral and select host transcripts, thereby facilitating viral replication and contributing to host shutoff. In line with these documented virus–spliceosome interactions, our findings are consistent with the notion that the host splicing machinery is involved in FHV-1 infection. Nevertheless, a direct functional characterization of Ess-2 in FHV-1 replication and host interaction remains essential. Beyond Ess-2, another key regulator of splicing is the kinase CDK13. CDK13 controls cell proliferation and differentiation by regulating key steps in gene expression [33]. Specifically, the RS motif in its N-terminal domain mediates interactions with spliceosomal components [34]. Furthermore, CDK13 controls splicing efficiency and fidelity by phosphorylating key splicing factors, thereby modulating their activity, localization, and interactions. However, it remains unknown whether CDK13 phosphorylation is functionally relevant to FHV-1 infection.

STAT5A transduces extracellular signals to the nucleus and directly regulates the expression of specific target genes. This transcriptional activation is strictly dependent on phosphorylation at two critical residues: Tyr694 and Ser726. Phosphorylation of Tyr694 enables STAT5A to translocate into the nucleus and bind DNA [35]. Phosphorylation at Ser726 further amplifies the transcriptional activity of STAT5A, ensuring robust expression of its target genes [36]. In the absence of Ser726 phosphorylation, STAT5A's ability to drive gene transcription is significantly compromised, even when Tyr694 is phosphorylated. BCLAF1 (Bcl-2-associated transcription factor 1) is a multifunctional nuclear protein that was originally identified through its interaction with the adenoviral Bcl-2 homolog E1B19K [37]. Subsequent studies have revealed the extensive roles of BCLAF1 in various processes, ranging from the DNA damage response (DDR), pre-mRNA splicing, and T cell activation to lung development, muscle cell proliferation/differentiation, autophagy, ischemia–reperfusion injury, and viral infection [38]. Moreover, emerging evidence identifies Bclaf1 as a host factor involved in anti-herpesviral defense, which is strikingly targeted by multiple viral components. The betaherpesvirus human cytomegalovirus (HCMV) employs multiple viral components—including the proteins pp71 and UL35, as well as a microRNA—to reduce cellular Bclaf1 levels [39]. However, whether Bclaf1 plays a role in alphaherpesvirus infection and the molecular mechanism of its antiviral activity remains to be elucidated.

In contrast to the pK$^-$ mutant, which primarily impacted the 'Spliceosome' and 'DNA replication', the TK$^-$ mutant was more strongly associated with the activation of pathways including 'ATP-dependent chromatin remodeling', 'Wnt signaling', and 'Neurotrophin signaling'. This indicates that distinct viral mutations influence the host through different molecular mechanisms. Furthermore, a group of molecules—including SMARCA5 and RSF1—which are linked to the ATP-dependent chromatin remodeling pathway and exhibit altered phosphorylation, merits future study.

SMARCA5 serves as the catalytic ATPase subunit in the ISWI family of ATP-dependent chromatin remodeling complexes. It utilizes energy from ATP hydrolysis to mediate short-range nucleosome sliding along DNA [40]. During infection by large DNA viruses such as HCMV and adenovirus, SMARCA5 is recruited to viral replication centers, where it facilitates viral genome replication—likely by maintaining chromatin structure around replication forks through its nucleosome remodeling activities [41,42]. During infection by viruses like HSV-1 or IAV, nuclear phosphorylated RPSA recruits SMARCA5, a process that enhances chromatin accessibility at NF-κB target gene promoters and facilitates the transcription of pro-inflammatory cytokines [43]. The RSF complex, an ISWI-family chromatin remodeler, uses energy derived from ATP hydrolysis—catalyzed by its SNF2H subunit—to slide nucleosomes along DNA [44]. This repositioning alters nucleosome arrangement while preserving their structural integrity. The RSF complex possesses ATP-dependent chromatin remodeling activity. It slides nucleosomes to expose or mask DNA regions, thereby regulating transcription factor binding and gene expression [45]. During HPV infection, the RSF complex (comprising RSF1 and SNF2H) is recruited by the viral E2 protein to participate in viral chromatin regulation, thereby potentially influencing the balance between viral gene expression and silencing [46]. Therefore, further investigation into the relationship between FHV-1 replication and the RSF complex is warranted to advance our understanding of viral replication mechanisms

Although our study focused on F81 cells infected with various FHV-1 mutants, the vast majority of identified proteins exhibit highly conserved sequences and functions across species. Thus, these findings may be broadly relevant to phosphorylation studies in other mammalian cells infected with related viruses.

Supplementary Materials: The following supporting information can be downloaded at: https://www.mdpi.com/article/10.3390/vetsci12121185/s1, Figures S1 and S2: Original PCR/Gel Electrophoresis pictures.

Author Contributions: Conceptualization: W.L., Y.L. and J.L., Methodology: W.L., X.Z., H.K. and H.J., Investigation: W.L., Q.J. and J.L. Visualization: H.K. and Q.J. Supervision: Y.L. and J.L. Writing—original draft: W.L., X.Z., J.L. Writing—review & editing: H.J., J.L. and Y.L. All authors have read and agreed to the published version of the manuscript.

Funding: This research received no external funding.

Institutional Review Board Statement: Not applicable.

Informed Consent Statement: Not applicable.

Data Availability Statement: The original contributions presented in this study are included in the article/Supplementary Material. Further inquiries can be directed to the corresponding author(s).

Conflicts of Interest: The authors declare that they have no competing interests.

References

1. Edwards, D.S.; Coyne, K.; Dawson, S.; Gaskell, R.M.; Henley, W.E.; Rogers, K.; Wood, J.L. Risk factors for time to diagnosis of feline upper respiratory tract disease in UK animal adoption shelters. *Prev. Vet. Med.* **2008**, *87*, 327–339. [CrossRef] [PubMed]
2. Davison, A.J.; Eberle, R.; Ehlers, B.; Hayward, G.S.; McGeoch, D.J.; Minson, A.C.; Pellett, P.E.; Roizman, B.; Studdert, M.J.; Thiry, E. The order Herpesvirales. *Arch. Virol.* **2009**, *154*, 171–177. [CrossRef]
3. Gaskell, R.; Dawson, S.; Radford, A.; Thiry, E. Feline herpesvirus. *Vet. Res.* **2007**, *38*, 337–354. [CrossRef]
4. Rota, P.A.; Maes, R.K.; Ruyechan, W.T. Physical characterization of the genome of feline herpesvirus-1. *Virology* **1986**, *154*, 168–179. [CrossRef]
5. Albà, M.M.; Das, R.; Orengo, C.A.; Kellam, P. Genomewide function conservation and phylogeny in the Herpesviridae. *Genome Res.* **2001**, *11*, 43–54. [CrossRef]
6. Tai, S.H.; Niikura, M.; Cheng, H.H.; Kruger, J.M.; Wise, A.G.; Maes, R.K. Complete genomic sequence and an infectious BAC clone of feline herpesvirus-1 (FHV-1). *Virology* **2010**, *401*, 215–227. [CrossRef]
7. Xu, M.; Wang, Y.; Liu, Y.; Chen, S.; Zhu, L.; Tong, L.; Zheng, Y.; Osterrieder, N.; Zhang, C.; Wang, J. A Novel Strategy of US3 Codon De-Optimization for Construction of an Attenuated Pseudorabies Virus against High Virulent Chinese Pseudorabies Virus Variant. *Vaccines* **2023**, *11*, 1288. [CrossRef]
8. Tai, S.H.; Holz, C.; Engstrom, M.D.; Cheng, H.H.; Maes, R.K. In vitro characterization of felid herpesvirus 1 (FHV-1) mutants generated by recombineering in a recombinant BAC vector. *Virus Res.* **2016**, *221*, 15–22. [CrossRef]
9. Tian, J.; Liu, Y.; Liu, X.; Sun, X.; Zhang, J.; Qu, L. Feline Herpesvirus 1 US3 Blocks the Type I Interferon Signal Pathway by Targeting Interferon Regulatory Factor 3 Dimerization in a Kinase-Independent Manner. *J. Virol.* **2018**, *92*, e00047-18. [CrossRef] [PubMed]
10. Ferrari, M.; Gualandi, G.L.; Corradi, A.; Monaci, C.; Romanelli, M.G.; Tosi, G.; Cantoni, A.M. Experimental infection of pigs with a thymidine kinase negative strain of pseudorabies virus. *Comp. Immunol. Microbiol. Infect. Dis.* **1998**, *21*, 291–303. [CrossRef] [PubMed]
11. Ferrari, M.; Mettenleiter, T.C.; Romanelli, M.G.; Cabassi, E.; Corradi, A.; Dal Mas, N.; Silini, R. A comparative study of pseudorabies virus (PRV) strains with defects in thymidine kinase and glycoprotein genes. *J. Comp. Pathol.* **2000**, *123*, 152–163. [CrossRef]
12. Cohen, P. The role of protein phosphorylation in human health and disease. The Sir Hans Krebs Medal Lecture. *Eur. J. Biochem.* **2001**, *268*, 5001–5010. [CrossRef]
13. Jensen, O.N. Interpreting the protein language using proteomics. *Nat. Rev. Mol. Cell Biol.* **2006**, *7*, 391–403. [CrossRef]
14. Pinna, L.A.; Ruzzene, M. How do protein kinases recognize their substrates? *Biochim. Biophys. Acta* **1996**, *1314*, 191–225. [CrossRef] [PubMed]
15. Caenepeel, S.; Charydczak, G.; Sudarsanam, S.; Hunter, T.; Manning, G. The mouse kinome: Discovery and comparative genomics of all mouse protein kinases. *Proc. Natl. Acad. Sci. USA* **2004**, *101*, 11707–11712. [CrossRef] [PubMed]
16. Gong, T.; Jiang, W.; Zhou, R. Control of Inflammasome Activation by Phosphorylation. *Trends Biochem. Sci.* **2018**, *43*, 685–699. [CrossRef]
17. Chattopadhyay, S.; Mukherjee, A.; Patra, U.; Bhowmick, R.; Basak, T.; Sengupta, S.; Chawla-Sarkar, M. Tyrosine phosphorylation modulates mitochondrial chaperonin Hsp60 and delays rotavirus NSP4-mediated apoptotic signaling in host cells. *Cell Microbiol.* **2017**, *19*, e12670. [CrossRef]
18. Huang, F.; Wu, Y.; Tan, H.; Guo, T.; Zhang, K.; Li, D.; Tong, Z. Phosphorylation of nucleolin is indispensable to its involvement in the proliferation and migration of non-small cell lung cancer cells. *Oncol. Rep.* **2019**, *41*, 590–598. [CrossRef]
19. Seo, J.; Guk, G.; Park, S.H.; Jeong, M.H.; Jeong, J.H.; Yoon, H.G.; Choi, K.C. Tyrosine phosphorylation of HDAC3 by Src kinase mediates proliferation of HER2-positive breast cancer cells. *J. Cell. Physiol.* **2019**, *234*, 6428–6436. [CrossRef] [PubMed]
20. Öhman, T.; Söderholm, S.; Paidikondala, M.; Lietzén, N.; Matikainen, S.; Nyman, T.A. Phosphoproteome characterization reveals that Sendai virus infection activates mTOR signaling in human epithelial cells. *Proteomics* **2015**, *15*, 2087–2097. [CrossRef]
21. Jansens, R.J.J.; Verhamme, R.; Mirza, A.H.; Olarerin-George, A.; Van Waesberghe, C.; Jaffrey, S.R.; Favoreel, H.W. Alphaherpesvirus US3 protein-mediated inhibition of the m6A mRNA methyltransferase complex. *Cell Rep.* **2022**, *40*, 111107. [CrossRef]
22. Liao, Y.; Lupiani, B.; Ai-Mahmood, M.; Reddy, S.M. Marek's disease virus US3 protein kinase phosphorylates chicken HDAC 1 and 2 and regulates viral replication and pathogenesis. *PLoS Pathog.* **2021**, *17*, e1009307. [CrossRef]
23. Stagni, V.; Orecchia, S.; Mignini, L.; Beji, S.; Antonioni, A.; Caggiano, C.; Barilà, D.; Bielli, P.; Sette, C. DNA Damage Regulates the Functions of the RNA Binding Protein Sam68 through ATM-Dependent Phosphorylation. *Cancers* **2022**, *14*, 3847. [CrossRef]

24. Padilla-Benavides, T.; Reyes-Gutierrez, P.; Imbalzano, A.N. Regulation of the Mammalian SWI/SNF Family of Chromatin Remodeling Enzymes by Phosphorylation during Myogenesis. *Biology* **2020**, *9*, 152. [CrossRef]
25. Lu, F.; Zhou, J.; Wiedmer, A.; Madden, K.; Yuan, Y.; Lieberman, P.M. Chromatin remodeling of the Kaposi's sarcoma-associated herpesvirus ORF50 promoter correlates with reactivation from latency. *J. Virol.* **2003**, *77*, 11425–11435. [CrossRef]
26. Musselman, C.A.; Kutateladze, T.G. Handpicking epigenetic marks with PHD fingers. *Nucleic Acids Res.* **2011**, *39*, 9061–9071. [CrossRef] [PubMed]
27. Tang, A.; Zhu, M.; Zhu, J.; Zhang, D.; Zhu, S.; Wang, X.; Meng, C.; Li, C.; Liu, G. Pathogenicity and immunogenicity of gI/gE/TK-gene-deleted Felid herpesvirus 1 variants in cats. *Virol. J.* **2023**, *20*, 87. [CrossRef] [PubMed]
28. Takada, I.; Hidano, S.; Nakagawa, T.; Nakagawa, S.; Makishima, M.; Takahashi, S. The Role of ESS2/DGCR14: Is It an Essential Factor in Splicing and Transcription? *Int. J. Mol. Sci.* **2025**, *26*, 4056. [CrossRef] [PubMed]
29. Prescott, E.L.; Brimacombe, C.L.; Hartley, M.; Bell, I.; Graham, S.; Roberts, S. Human papillomavirus type 1 E1^E4 protein is a potent inhibitor of the serine-arginine (SR) protein kinase SRPK1 and inhibits phosphorylation of host SR proteins and of the viral transcription and replication regulator E2. *J. Virol.* **2014**, *88*, 12599–12611. [CrossRef]
30. Dhillon, P.; Tandra, V.N.; Chorghade, S.G.; Namsa, N.D.; Sahoo, L.; Rao, C.D. Cytoplasmic Relocalization and Colocalization with Viroplasms of Host Cell Proteins, and Their Role in Rotavirus Infection. *J. Virol.* **2018**, *92*, e00612-18. [CrossRef]
31. Thompson, M.G.; Muñoz-Moreno, R.; Bhat, P.; Roytenberg, R.; Lindberg, J.; Gazzara, M.R.; Mallory, M.J.; Zhang, K.; García-Sastre, A.; Fontoura, B.M.A.; et al. Co-regulatory activity of hnRNP K and NS1-BP in influenza and human mRNA splicing. *Nat. Commun.* **2018**, *9*, 2407. [CrossRef]
32. Bryant, H.E.; Wadd, S.E.; Lamond, A.I.; Silverstein, S.J.; Clements, J.B. Herpes simplex virus IE63 (ICP27) protein interacts with spliceosome-associated protein 145 and inhibits splicing prior to the first catalytic step. *J. Virol.* **2001**, *75*, 4376–4385. [CrossRef]
33. Pitolli, C.; Marini, A.; Sette, C.; Pagliarini, V. Physiological and pathological roles of the transcriptional kinases CDK12 and CDK13 in the central nervous system. *Cell Death Differ.* **2025**, *32*, 371–381. [CrossRef]
34. Cho, S.; Hoang, A.; Sinha, R.; Zhong, X.Y.; Fu, X.D.; Krainer, A.R.; Ghosh, G. Interaction between the RNA binding domains of Ser-Arg splicing factor 1 and U1-70K snRNP protein determines early spliceosome assembly. *Proc. Natl. Acad. Sci. USA* **2011**, *108*, 8233–8238. [CrossRef] [PubMed]
35. Gouilleux, F.; Wakao, H.; Mundt, M.; Groner, B. Prolactin induces phosphorylation of Tyr694 of Stat5 (MGF), a prerequisite for DNA binding and induction of transcription. *Embo J.* **1994**, *13*, 4361–4369. [CrossRef] [PubMed]
36. Benitah, S.A.; Valerón, P.F.; Rui, H.; Lacal, J.C. STAT5a activation mediates the epithelial to mesenchymal transition induced by oncogenic RhoA. *Mol. Biol. Cell* **2003**, *14*, 40–53. [CrossRef] [PubMed]
37. Kasof, G.M.; Goyal, L.; White, E. Btf, a novel death-promoting transcriptional repressor that interacts with Bcl-2-related proteins. *Mol. Cell. Biol.* **1999**, *19*, 4390–4404. [CrossRef]
38. Yu, Z.; Zhu, J.; Wang, H.; Li, H.; Jin, X. Function of BCLAF1 in human disease. *Oncol. Lett.* **2022**, *23*, 58. [CrossRef]
39. Lee, S.H.; Kalejta, R.F.; Kerry, J.; Semmes, O.J.; O'Connor, C.M.; Khan, Z.; Garcia, B.A.; Shenk, T.; Murphy, E. BclAF1 restriction factor is neutralized by proteasomal degradation and microRNA repression during human cytomegalovirus infection. *Proc. Natl. Acad. Sci. USA* **2012**, *109*, 9575–9580. [CrossRef]
40. Aydin, Ö.Z.; Marteijn, J.A.; Ribeiro-Silva, C.; Rodríguez López, A.; Wijgers, N.; Smeenk, G.; van Attikum, H.; Poot, R.A.; Vermeulen, W.; Lans, H. Human ISWI complexes are targeted by SMARCA5 ATPase and SLIDE domains to help resolve lesion-stalled transcription. *Nucleic Acids Res.* **2014**, *42*, 8473–8485. [CrossRef]
41. Bryant, K.F.; Colgrove, R.C.; Knipe, D.M. Cellular SNF2H chromatin-remodeling factor promotes herpes simplex virus 1 immediate-early gene expression and replication. *mBio* **2011**, *2*, e00330-10. [CrossRef]
42. Brestovitsky, A.; Sharf, R.; Mittelman, K.; Kleinberger, T. The adenovirus E4orf4 protein targets PP2A to the ACF chromatin-remodeling factor and induces cell death through regulation of SNF2h-containing complexes. *Nucleic Acids Res.* **2011**, *39*, 6414–6427. [CrossRef]
43. Jiang, Y.; Sun, S.; Quan, Y.; Wang, X.; You, Y.; Zhang, X.; Zhang, Y.; Liu, Y.; Wang, B.; Xu, H.; et al. Nuclear RPSA senses viral nucleic acids to promote the innate inflammatory response. *Nat. Commun.* **2023**, *14*, 8455. [CrossRef] [PubMed]
44. Zhang, J.; Zhao, H.; Zou, B.; Li, H.; Dong, S.; Guan, J.; Wang, C.; Li, W.; Liu, Y.; Chen, Y.; et al. Cryo-EM structure and functional analysis of the chromatin remodeler RSF. *Acta Crystallogr. F Struct. Biol. Commun.* **2024**, *80*, 125–134. [CrossRef] [PubMed]
45. Hanai, K.; Furuhashi, H.; Yamamoto, T.; Akasaka, K.; Hirose, S. RSF governs silent chromatin formation via histone H2Av replacement. *PLoS Genet.* **2008**, *4*, e1000011. [CrossRef] [PubMed]
46. Muller, M.; Demeret, C. The HPV E2-Host Protein-Protein Interactions: A Complex Hijacking of the Cellular Network. *Open Virol. J.* **2012**, *6*, 173–189. [CrossRef]

veterinary sciences

MDPI

Article

Orthodontic Implants: Novelty and Evolution in Veterinary Orthodontics—Retrospective Case Series Report

Raluca-Ioana Nedelea [1,*,†], Mihai Marian Borzan [2,†], Cristinel Cezar Mătură [3] and Ioan Marcus [1]

[1] Pathophysiology Department, University of Agricultural Studies and Veterinary Medicine, 400372 Cluj-Napoca, Romania

[2] Animal Breeding and Animal Productions Department, University of Agricultural Studies and Veterinary Medicine, 400372 Cluj-Napoca, Romania

[3] DI-VET Medical, 011635 Bucharest, Romania

* Correspondence: raluca-ioana.pop@student.usamvcluj.ro

† These authors contributed equally to this work.

Simple Summary: Orthodontic implants (OIs) are titanium screws inserted into the maxilla or mandible to serve as anchorage for moving teeth, which are often confused with dental implants. This article will provide a comprehensive comparison of the two types of screws used in dentistry, and three representative cases illustrating indications for the use of OIs in veterinary dentistry. Frequently, OIs are used to move the lower incisors lingually or to move the upper canines backwards. Moreover, their use extends to palate surgery, with screws used to hold a custom-made shield in place. This article will cover topics regarding the selection criteria, the best sites for insertion, the techniques for insertion and removal, and their possible complications, along with the prevention and management of said complications.

Abstract: Orthodontic implants (OIs) are endosseous devices designed to provide stable skeletal anchorage in orthodontic treatment. A variety of terms have been proposed for these devices; however, such terminology may lead to OIs being confused with conventional dental implants. Therefore, the similarities and distinctions between OIs and dental implants will be examined in this research. Three representative clinical cases will be presented, illustrating the application of OIs for mandibular incisor teeth lingualizations, distalization of a maxillary canine tooth, and palatal flap protection for palatoschizis closure in a cat, describing a custom-made protection shield held in place with four OIs. This article will cover the criteria for OI selection, the determination of optimal insertion sites, and the placement technique. Given that a key prerequisite for a successful outcome is primary stability, several complications may occur such as local inflammation, injury to adjacent anatomical structures, implant fracture, loss of stability, or challenges in attaching elastic chains due to interference from surrounding soft tissues. The prevention and management of such complications will be addressed.

Keywords: veterinary orthodontics; orthodontic implants; dogs; cats; veterinary dentistry

1. Introduction

Malocclusions in veterinary dentistry are frequently encountered [1–4], and owners are increasingly requesting orthodontic treatments. Head conformation impairments in dogs can be classified into three groups: mesocephalic, dolichocephalic, and brachycephalic [5]. The mesocephalic conformation forms the reference group, as this conformation is encountered in wildlife and is considered normal, even in humans. Dolichocephalic heads are long and narrow, as in Greyhounds or Border Collies, and brachycephalic breeds, the most representative being the French Bulldog, typically have short and broad muzzles, and often exhibit a high incidence of dental pathologies. In mesocephalic dogs, the maxilla is slightly larger than the mandible, with the maxillary incisor teeth overlapping the mandibular ones. Other occlusal keys are present in canines, where the tips of mandibular canine teeth are located midway between the third incisor tooth and the maxillary canine tooth. The maxillary premolars are buccally situated, presenting interdigitation, with the maxillary teeth occluding rostrally to the mandibular ones. The maxillary fourth premolar occludes on the buccal side of the mesial cusp of the mandibular first molar, and the molars have occlusal surfaces, which favors the crushing of food [5]. Human dentistry recognizes three classes of malocclusion, as classified by Edward Hartley Angle, who is considered the father of modern human orthodontics [6]. Class I refers to eugnathic occlusion, described as the reference occlusion in mesocephalic heads, and refers to a mesiodistal relationship between the first molars in humans, or carnassials in veterinary dentistry. Malocclusions occur in varying forms; thus, Dewey [7] and Anderson [8] further categorized Class I into five types of malocclusions:

1. Dewey type 1—incisors crowded and/or canines lingually positioned;
2. Dewey type 2—incisors protruding;
3. Dewey type 3—edge-to-edge contact and/or anterior crossbite;
4. Anderson type 4—uni-/bilateral posterior crossbite;
5. Anderson type 5—mesioversion of the first permanent molar (the occlusal key for human orthodontics) due to premature extraction of deciduous teeth.

Class II refers to mandibular distocclusion, where the mandible is retrognathically positioned to the maxilla [9]. Class II is also divided into two subclasses:

1. Type 1—the maxillary central incisors are palatally inclined and may be overlapped by the maxillary lateral incisors.
2. Type 2—a deep overbite and a broad maxillary arch.

Class III refers to mandibular mesiocclusion, where the mandible is situated anterior to the maxilla [10], and is divided into three classes:

1. Type 1—the dental arch is abnormally shaped;
2. Type 2—the mandibular teeth are lingually oriented;
3. Type 3—the maxillary teeth are palatally oriented palatally

For veterinary dentistry, the AVDC has established categories for symmetrical and asymmetrical skeletal malocclusions [11]. There are three classes for symmetrical skeletal malocclusion: Class I, referring to eugnathic occlusion, but with some teeth rotated or malpositioned, Class II, referring to mandibular distocclusion; and Class III, referring to mandibular mesiocclusion. In cases of no apparent discrepancy between the jaws, the symmetrical skeletal malocclusions in these three classes are known as dental malocclusions. Asymmetrical skeletal classes are recognized under Class IV malocclusions and include maxilla–mandibular asymmetry in the rosto-caudal, side-to-side, or dorso-ventral directions.

Orthodontics in veterinary dentistry typically addresses Class I dental malocclusions, characterized by malpositioned incisor and/or canine teeth. Conventional orthodontics involves the movement of teeth with the support of adjacent teeth or with the use of extraoral devices [12]. Extraoral anchorage has historical significance in human dentistry, but it is not applicable in veterinary dentistry. Every tooth movement requires an anchorage that enables correct tooth movement and must not be displaced during orthodontic treatment. In veterinary dentistry, however, the surgeon cannot control the behavior or biting force of the animal, so reliable approaches to achieve these goals are needed.

OIs are endosseous devices designed to provide stable skeletal anchorage in orthodontic treatment. Although they do not have a unique definition in the literature, they are synonymous with mini-screws, mini-implants, and micro-implants [13,14]. The ISO 16443 (Dentistry—Vocabulary of Oral Implantology) [15] specifies terms and definitions for dental implants, instruments, and accessories, as well as the most commonly used clinical terms in the field of dental implantology. According to its definitions, a DI, or dental implant, is defined as a device primarily designed to be placed within, through, or upon the bones of the craniofacial complex to support and to resist displacement of a dental prosthesis, replacing missing teeth. An OI, by contrast, is a device specially designed to be placed through the bones of the craniofacial complex with the primary purpose of providing anchorage for an orthodontic appliance [16]. Unlike conventional orthodontic mechanics, which depend on the support of adjacent dental units and may consequently result in undesired tooth movement or anchorage loss, OIs offer a direct and predictable alternative. Thus, local anatomical conditions accompanied by insufficient dental support for stable orthodontic anchorage are indications for the use of OIs. In human orthodontics, they offer a reliable anchor in edentulous areas or in cases of periodontal disease. In veterinary dentistry, periodontal disease is less likely to be present, as orthodontic treatment typically occurs at an early age. While the utility of DIs in small animals is still questioned, OIs are beginning to be used more frequently. Kanomi [17] designed titanium OIs with a diameter of 1.2 mm and a length of 6 mm, which first entered the market in 1997. Titanium was chosen to enhance the anchorage, as it was widely used for DIs due to its biocompatibility and ability to prevent infections [18].

Palatoschisis in cats and dogs is a life-threatening condition with a multifactorial etiology [5,19]. Several surgical techniques have been described for closing this palatal defect in correlation with the following dimensions: narrow congenital hard palate clefts may be rectified with the medially positioned flap technique [5,20,21], and for wider hard palate clefts, the overlapping flap technique ensures tension-free closing of the palatal defect [5,22,23]. The hard palate may either be addressed in one session together with the soft palate, or the two procedures may be undertaken separately. When the palate defect is wide, the lack of bony support may be challenging for flap positioning, threatening its survival. Moreover, as a cat's tongue has numerous filiform papillae [24], it may scrape the flap where there is no underlying bony support, which may compromise the surgical results. Given these aspects, the authors designed a customized shield to protect surgical flaps from abrasions caused by tongue papillae. This custom-made shield was held in place with the help of four OIs. The materials and methods will be described and discussed later in this article.

The literature on human orthodontics is abundant on this topic; however, veterinary dentistry lacks information. Thus, the aim of this article is to present retrospective case series with OIs' indications, techniques, characteristics, and complications—and their prevention and management—regarding the use of OIs in veterinary dentistry, placing it within the context of findings from human dentistry.

2. Materials and Methods

Twelve cases were selected in which OIs were used for orthodontic reasons in dogs, and one case involved the use of OIs to protect a flap used in palatoschisis correction in a cat. Inclusion criteria for the cases for malocclusion were class I malocclusion. All patients had permanent dentition, with ages ranging from 5.5 months to 26 months with a median age of 10.95 months. Gender was not considered a determining factor. Mandatory criteria for inclusion in the study were a proper medical condition to undergo safely multiple general anesthetics, the owner's willingness, as evidenced by signed informed consent, a discrepancy between the maxilla and mandible of less than 5 mm, and neutered patients. A thorough preoperative medical exam and preoperative laboratory tests, including hematology, serum biochemistry, and blood coagulation, confirmed a proper medical condition. Every selected orthodontic case had an ASA physical status of I. Premedication was performed with medetomidine (0.005 mg/kg i.m.) and methadone (0.5 mg/kg i.m.). Every patient, without exception, received an intravenous catheter through which propofol (2 mg/kg i.v.) was used to induce anesthesia. Endotracheal tubes were used to administer inhaled anesthetics. MAC for isoflurane was 1.3%. Monitoring during orthodontic procedures included evaluation of cardiac rate and rhythm, oxygenation, respiratory rate and rhythm, body temperature, anesthetic depth, capillary refill time, oral mucosa color, blood pressure, and analgesia. Local anesthesia was obtained with Lidocaine 2% (5 mg/kg) for a caudal mental nerve block for OIs applied on the mandible or an infraorbital nerve block for OIs applied on the maxilla.

Exclusion criteria were other malocclusions than rostral crossbite, rostral level bite, or mesialized maxillary canine tooth; ASA physical status other than I; non-cooperant owner; or refusal to neuter the patient.

In each case, extraoral photographs were captured from the frontal and both lateral views, along with intraoral photographs at the beginning of the treatment, during and after. The images were taken using a Huawei Mate 20 Pro, (Huawei Technologies Co., Ltd., Shenzhen, Guangdong, China) with the Macro function enabled, and a Smile Lite MDP1 (Smile line, St-Imer, Swizerland) circular light device, specifically designed for accurate dental photography. Orthodontic cases began with taking dental impressions of both dental arches using condensation silicones. The dental laboratory then scanned these impressions, leading to the creation of orthodontic devices through Computer-Aided Design (CAD) and Computer-Aided Manufacturing (CAM), Exocad 2023. To facilitate tooth movement, metallic crowns on the incisor teeth were designed with hooks on the lingual part and inclined planes on the incisal parts. These hooks were connected to the orthodontic implants using elastic chains, which needed to be changed every two weeks. At the end of the treatment, the orthodontic devices were cemented in place and removed using tungsten cylindrical burs and ultrasonic scalers.

Osstem produced the OIs used in the procedure and were manufactured in Busan, South Korea. The surgical kit utilized for implant insertion was provided by the manufacturer and included a pilot drill, hand drivers of varying lengths, handpiece drivers of different dimensions, and heads designed to fit various shapes of OIs.

All OIs were placed without angulation with the hand driver after pilot drilling. In eleven of these cases, OIs were used to lingualize the mandibular incisor teeth, and in one case, the OIs were used to distalize a maxillary canine tooth. All of the dogs included in this study were neutered. Eight cases presented a rostral cross-bite, and three cases presented a rostral level bite. Two cases, one for each indication, were selected to illustrate the use of OIs in dogs for orthodontic purposes, with the third case presenting the use of OIs for protecting the flap in palatoschisis management in a four-month-old cat, thus enlarging the range for OI indications.

OIs insertion involved the following steps:

1. Dental X-ray of the mandibular premolar teeth to determine the precise localization of the OI.
2. Regional nerve block.
3. Pilot drilling in cortical bone.
4. Dental X-ray.
5. Insertion of the OI.
6. Final dental X-ray.

Dental X-rays were obtained using phosphor plates, which allowed images to be transferred to a computer via a scanning device (indirect digital radiography). The size of the plates used was either 2 or 4, depending on the breed and dimensions of the head. Both parallel technique and bisecting angle technique were utilized during the X-ray process. Dental X-rays were taken to accurately determine the diameter of the OIs, localize them precisely, verify their correct placement, and ensure that the OIs did not interfere with the dental roots.

Every case was checked visually/taking pictures every two weeks, and if needed (more tooth movement required), elastic chains were changed under general anesthesia. A four-month-old cat was presented with a wide congenital cleft of the hard palate (palatoschisis). Firstly, surgery was performed on the hard palate, and after six months it was performed on the soft palate. In the surgical approach for the hard palate, the overlapping technique was used. A dental impression was taken with condensation silicone. A model of polymethylmethacrylate (PMMA) was made immediately and isolated with medical Vaseline. On the model, a 1:1 PMMA plaque was made on the entire hard palate. The protection shield was applied at a distance of approximately 2 mm from the flap to avoid pressure to the flap, and it was carefully polished. Four holes were made in the corners using a diamond bur with a diameter of 1.4 mm, allowing for OI placement. OIs that were 1.4 mm in diameter and 10 mm in length were inserted to protect the flap. As the OI heads were button-headed, the surgeons applied a colored photopolymerized composite on them in order to avoid hurting the tongue and to allow regular feeding. The OIs with the PMMA plaque were removed two weeks later.

OI removal is a straightforward procedure, made under general anesthesia, and does not require local anesthesia. OIs are removed by twisting them in the direction opposite to insertion with the same screw that were inserted.

3. Results

3.1. Case 1—Lingualization of Mandibular Incisor Teeth

A small-breed dog, aged five and a half months, was presented for a rostral crossbite. Preoperative laboratory tests and the medical examination revealed no contraindications for undergoing multiple general anesthetics. A classic dental impression of the dental arches was made using condensation silicones. The dental laboratory scanned the impression and proceeded with the CAD/CAM of the future orthodontic device. Both the mandibular and maxillary incisor teeth had metallic crowns with inclined planes on the incisal edges. The mandibular crowns were designed with lingual hooks to facilitate the attachment of elastic chains. The metallic crowns were cemented onto the teeth without any mechanical or chemical surface treatment—only simple drying of the crowns was performed. Two button-head orthodontic implants (OIs) were placed buccally between the roots of the mandibular fourth premolars (308 and 408). The diameter of the OIs was determined by subtracting 2 mm from both sides of the dental roots at the chosen localization between the roots. Measurements were taken from the tip of the premolar's median cusp. For extra-small breeds, a minimal diameter of 1.2 mm is recommended, with dimensions increasing according to available bone. For a diameter of 1.4 mm, a minimum of 5.4 mm between the dental roots is required for proper insertion without risking their damage. In this case, the OIs measured 1.4 mm in diameter and 6 mm in length. They were inserted perpendicular to the mandible, without angulation. The jaw width was measured with a device to estimate the required length. After pre-drilling with the appropriate hand driver for button heads, the OIs were inserted to achieve primary stability, ensuring they were not too close to the cortical bone or exerting pressure on the soft tissues.

A two-week follow-up showed the occurrence of the occlusal jump and the correction of the rostral crossbite, with no need to change the elastic chains. This arrangement was maintained for two months to stabilize the results. At the end of the maintenance period, the orthodontic device was removed. A follow-up after two years revealed that the results were stable, with no changes in the color or morphology of the crown surfaces. Some aspects described are illustrated in Figure 1.

Some of the aspects described are illustrated in Figure 1.

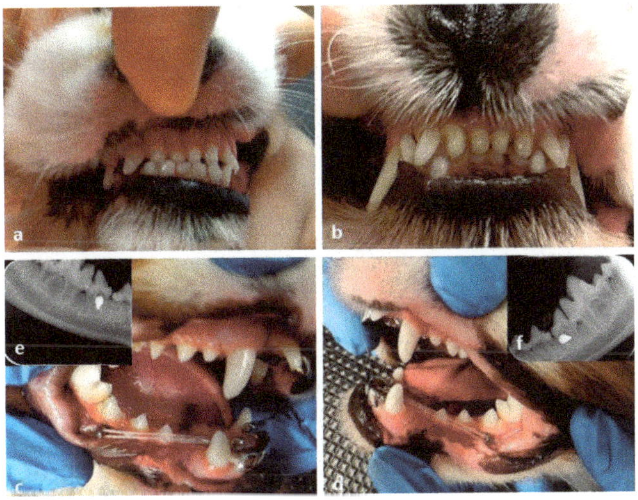

Figure 1. (**a**) Clinical view of the rostral crossbite. Note the inversed occlusion between 102, 101, 201, 202 and 302, 301, 401, 402, 403. The dental units were in contact. (**b**) Clinical view at the end of the orthodontic treatment. (**c**) Clinical view of the elastic chains attached to the OI—right side view. Note the lingual hooks from the lingual side. (**d**) Clinical view of the elastic chains attached to the OI—left side view. Note the lingual hooks attached to the button head OIs. (**e**) Radiological view of the inserted OI—right side view. (**f**) Radiological view of the inserted OI—left side view.

3.2. Case 2—Mesially Displaced Canine Tooth in a Medium-Size-Breed Dog

The dog exhibited lingualization of tooth 304, along with mesialization of tooth 204. In this case, considering the size of the canine root, two orthodontic implants (OIs) were placed: one in the zygomatic arch, parallel to tooth 208, and another between teeth 207 and 208 along the dental arch. The maxillary canine had a metallic collar that was one centimeter high, featuring two buttons on the labial side and one hook on the distal side of the crown. The buttons connected to the OI from the zygomatic arch, while the hook linked to the OI situated between teeth 207 and 208, using elastic chains. These elastic chains were changed every two weeks, three times. The mandibular canine required labioversion, so an orthodontic device was designed to apply forces from the lingual side to the labial side. This device in the mandibular arch consisted of two metallic collars on the mandibular canines, connected to a lingual component that included an orthodontic expanding screw. The expanding screw was activated each time the elastic chains were changed. Once there was sufficient space for the tip of the crown of tooth 304 to be positioned between teeth 203 and 204, through the distalization of tooth 204 and the labioversion of tooth 304, the occlusion of tooth 304 was normalized. All our orthodontic cases had maintenance interval of eight weeks. Several key aspects of this case are illustrated in Figure 2.

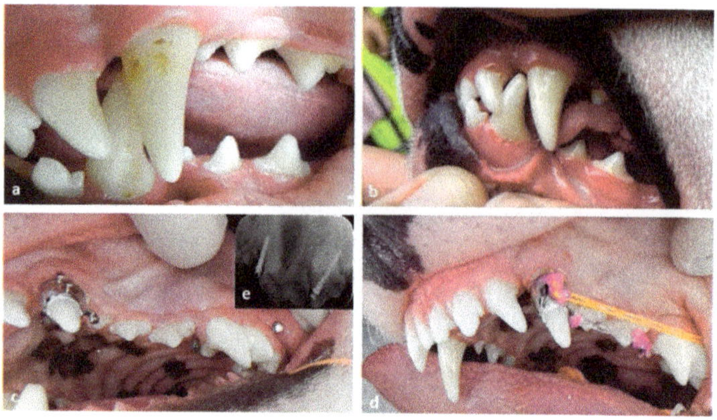

Figure 2. (**a**) Clinical view of the narrowed space between 203 and 204, and palatally displaced 304, impinging the palatal mucosa. (**b**) Clinical view at the end of the orthodontic treatment. (**c**) Clinical view of the orthodontic devices: a collar with two labial buttons and a distal hook, placed on the 204 and two OIs—one mesial and one buccally placed to 208. (**d**) Elastic chain placed on a labial button on 204 and colored light cured composite placed on the buttons to avoid trauma of the soft tissues. (**e**) Dental X-ray of the two OIs.

3.3. *Case 3—Congenital Cleft Palate in a Cat*

A four-month-old domestic short-haired cat was presented with large palatoschisis. Given the anatomy of a cat's tongue, the primary concern was the post-surgery period and the abrasive effect of the tongue that would impinge on the healing process by compromising the flap. The surgical team designed a custom-made protection shield made of PMMA to be applied for two weeks, enabling primary healing of the flap. The protection shield was placed 2 mm from the flap to prevent pressure on the flap and necrosis. It was held in place for two weeks with the help of four OIs. The OIs were selected, and the 10 mm length was calculated by taking into account several factors: the width of the flap, the distance from the shield to the flap, the shield's width, and a minimal anchor on the palatal bone. At two weeks post-surgery, the custom-made shield was removed, revealing granulation tissue, and the intact sutures. There was no presence of necrotic tissue. By three months after the surgery, the palatal mucosa was keratinized and fully healed. A two-year follow-up confirmed complete closure of the hard palate. The confirmation was obtained clinically and through a computed tomography scan. Although the scan was initially obtained to assess an inner ear pathology, it was also shared with the maxillofacial team for further evaluation. Several key aspects of the case are presented in Figure 3.

3.4. *General Results*

All OIs constituted a reliable anchorage system, facilitating precise and controlled orthodontic tooth movement. None of the cases presented infection or mobility issues, and primary stability was achieved in all cases. OIs that were used to maintain the results in the palate surgery remained in place, offering good stability for the protection shield.

None of the cases at the end of the orthodontic treatment and at two-year follow-up point showed pink or any other staining. Intra-vitam pink teeth were not associated with the use of OIs.

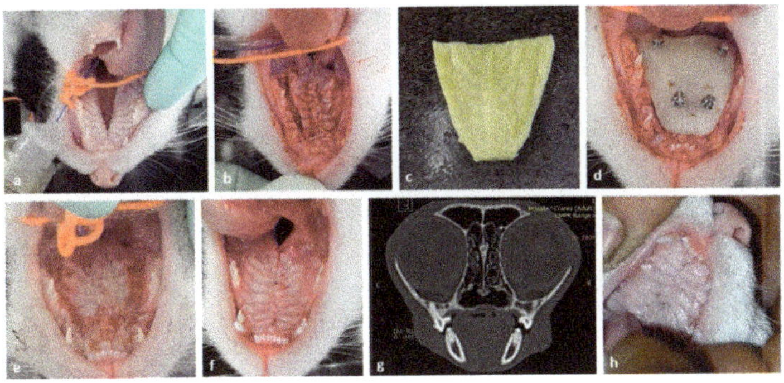

Figure 3. (**a**) Clinical view of the cleft palate in a four-month-old domestic short-haired cat. Note the wide palatal defect. (**b**) Intraoral view of the cleft palate treated using the overlapping technique. (**c**) Impression made with putty condensation silicone. (**d**) Protection shield applied intraorally with four OIs. (**e**) Clinical view of the surgical field at two weeks from after palate surgery. (**f**) Clinical view of the palate three months after surgery. (**g**) CT image of the hard palate two years after surgery. (**h**) Clinical view of the palate two years after surgery.

4. Discussion

The three cases presented summarize the primary indications for OIs in veterinary dentistry.

An OI has an endosseous part, a head, a transmucosal part, a body, and a tip. Classification of OIs is made regarding head design (button vs. bracket), diameter (varying from 1.2 to 2 mm), length (5 to 12 mm), body design (cylindric or tapered), transmucosal profile (from 0 to 3 mm), insertion technique (self-drilling and pre-drilling), thread orientation (clockwise or anti-clockwise), the alloy used for manufacturing them (titanium alloy or stainless steel), and sterility (sterilized or non-sterilized) [13]. The OIs used for obtaining the needed results were made of titanium, were sterilized, were button-headed, had a clockwise thread orientation, and had a transmucosal profile measuring 2 mm. The choice of design for the head was based on its scope, to ensure correct placement of the elastic chains. Moreover, the button head design was chosen as it is less traumatic to nearby mucosa. The manufacturer offered other designs, which ultimately were not chosen.

When choosing the OI, one should consider bone availability and the presence of a two-millimeter distance to nearby dental roots [25]. Dental X-rays taken before, during, and after OI insertion are taken to provide their precise location, indicating insertion sites at a distance from anatomical structures that would be damaged if OIs were inserted in them. The most commonly used sites in veterinary orthodontics are situated between the mandibular premolar roots, in the buccal side. If OIs are inserted lingually, not only is the OI insertion technique highly complex but also traumatic injury of the tongue and underlying tissues can occur. The length and diameter of an OI must correspond to the available bone volume. Thus, smaller breeds require OIs of smaller lengths and diameters than larger breeds, and each case should be approached individually.

As OIs are sometimes confused with DIs and lack precise differentiation, the authors have synthesized the differences between these two dental devices in Table 1 [15,26–28].

Primary stability (PS) in OIs refers to mechanical stability immediately after OI insertion. OIs are only measured via PS, by attempting to move the OI laterally with a tweezer, during which the OI should not exhibit any movement. By contrast, DIs are measured via the Implant Stability Quotient (ISQ), which can be measured at every moment of a DI's life [4,29,30]. PS depends on various factors, including bone density, cortical width, thread design, and diameter and length of the OI, as well as a correct and clean insertion technique [16]. If the OI presents mobility, it should be removed, as it does not offer reliable stability for tooth movement. Inserting the OI perpendicularly into the cortical bone usually ensures the PS is needed.

Table 1. The main differences between OIs and DIs.

Characteristics	OI	DI
1. Purpose	Anchorage	Replacing missing tooth
2. Time in the oral cavity	Limited period of time/months	Extended period of time/years
3. Osteointegration	Not necessary	Mandatory
4. Dimensions	Smaller	Larger
5. Diameter	1.2–2.0 mm	3 to 7 mm or more
6. Length	5–12 mm	6–25 mm
7. Coronal part	Button/Bracket	Cylindrical/conical form
8. Localization	Outside the dental arch	On the dental arch
9. Associated inflammation	Minimal local inflammation	Periimplantitis
10. Stability measurement	Approximation	Precise measurement/ISQ
11. Types of stability needed	Primary stability needed	Every moment of a DI's life needed
12. Acceptance in veterinary dentistry	Accepted	Questionable
13. Surface treatment	None	Yes
14. Price	Low	High

Increasing the torsional force during OI insertion may lead to bending or fracture of the OI [16]. Kim et al. [31] investigated the issue of whether to insert OIs via drilling or not, as well as the relationships between OI dimensions and PS, and long-term results and complications. Even though PS is increased if an OI is inserted without drilling, studies recommend the use of a pilot drill [32]. The risk of OI fracture is very high if inserted without pre-drilling, with OIs usually fracturing at the least resistant part, at the joint between the head and body in the cervical region [32]. If the site is not predrilled, the OI can be displaced into the soft tissues, which complicates its removal. When predrilling, it is necessary to drill in a steady, correct direction, which is, in a great majority of cases, perpendicular to the cortical bone. Otherwise, if the hand moves while drilling, the resulting cavity will be larger, compromising PS when inserting the OI. Pilot drilling is also reported to reduce microdamage to the cortical bone [15,32].

Additionally, excessive OI loading can induce osteolysis and lead to OI mobility.

Another essential aspect is the placement level of the OI head. It is necessary that the head remains above the soft tissue and does not penetrate the bone or come into close contact with it. Otherwise, if these complications occur, it then becomes difficult to insert elastic chains and keep them in place. Regarding the distance from the nearby teeth, 2 mm of bone should be present between the OI and the teeth [31].

Removing OIs is not a complicated procedure; however, the surgeon must know the orientation of the threads, as there are different systems containing threads oriented to the right or to the left. Thus, the thread orientation should be noted in the medical records so that when removing the OIs, high force should not be applied, thus avoiding fractured OIs.

Management of a fractured OI typically involves incision and small-bone drilling around the OI to facilitate extraction. The authors do not recommend leaving a fractured OI in place, as it may cause a foreign body reaction.

Another complication that may arise with regard to OIs is mucosal inflammation. In our study, inflammation in cases of OIs in small animals was minimal and resolved after the OIs were removed.

Regarding the use of OIs in patients with periodontal pathology, one should pay attention to the force applied, as the alveolar bone is smaller in volume in patients affected by periodontal disease. Applying pressure to the affected periodontal ligaments and alveolar bone may lead to gingival recession, dental mobility, or even tooth loss [31]. It is important to note that in human dentistry, periodontal therapy precedes orthodontic treatment. Only after achieving the necessary health status of the periodontal structures and maintaining the results for a specified period is the orthodontic treatment initiated. If oral hygiene is poor and the results of periodontal therapy are weak, orthodontic treatment should not be initiated. However, in veterinary dentistry, this rule is hard to implement. The animal lifespan is shorter, and all procedures are performed under general anesthesia. If periodontal treatment with guided bone regeneration is performed in small animals before orthodontic treatment, the results remain questionable, and any periodontal problems are enhanced by abnormal forces applied to the periodontal ligaments due to malpositioned teeth. Plaque accumulation is favored by malpositioned teeth [33,34], making self-cleaning harder to manage. Nevertheless, the principle in human orthodontics to achieve periodontal stability before initiating orthodontic treatment [35] does not apply in veterinary orthodontics; thus, a small animal patient with periodontal issues is excluded from orthodontic treatment.

Other complications associated with OI insertion have been widely documented in human medicine [36]. One significant complication that can be prevented is damage to the dental roots. By carefully examining and calculating the dimensions via dental X-rays, one can insert an OI precisely without damaging the roots, thereby avoiding potential negative consequences in the long term. In human dentistry, pink staining due to internal resorption that occurs in conjunction with orthodontic treatment has been reported [37]. In veterinary dentistry, the appearance of an intra-vitam pink tooth was reported as a result of a DI insertion in a dog [38]. Human studies have documented that pink staining implies higher insertion forces when the OI is inserted into the roots compared to OI insertions not involving any root [39,40]. Radiographs taken before and during the procedure associated with the 2 mm rule can serve to ensure an insertion without damaging the nearby dental units. In our study, none of the cases treated with OIs at the end of the orthodontic treatment showed any staining of the dental crown.

Regarding the perforation of the nasal floor in the presented case of the cat, the healing was spontaneous without any oronasal fistula. The length of the OIs was 10 mm; the plate had a width of 2–3 mm and was placed at a distance of 2–3 mm from the flap. The flap was approximately 3 mm thick, and the nasal cavity was not perforated. Each OI was only inserted in the cortical bone of the palate, with the distribution of the four OIs ensuring stability of the protection shield. Similarly, human dentistry has reported spontaneous healing of the maxillary sinus in cases of perforation smaller than 2 mm [41].

A traumatic soft tissue injury to the buccal adjacent mucosa, which appeared as a kissing ulceration, was self-limiting and resolved by applying a colored-light-curing composite on the head of the OI without bonding. Thus, the composite was easily removed, and the head of the OI was not damaged during the change in elastic chains. This protocol, which utilizes light-curing materials, was successfully implemented in human dentistry [42].

Custom-made palatal shields have been used in human oro-maxillo-facial surgery to protect the newly restored palate and are maintained in place without any OI, by being applied and held in place by the adjacent teeth [43]. However, the risk of losing the shield by anchoring it to the cat's teeth was high, given that the cat was in the stage of transitioning from temporary to permanent teeth. Therefore, the authors considered maintaining the shield with the help of OIs. The OIs provided adequate protection, and healing occurred without any fistulae. A two-year follow-up of the case revealed the growth of the palatal bone, as seen in CT findings. Regarding the palatal growth that follows surgery, the authors report findings similar to those of Wright et al. [44], who reported spontaneous regeneration of a mandibular body after a mandibulectomy. The ages at which the surgeries were performed were similar, and the finding that the palatal bone continued to grow following cleft palate surgery confirms that even in the maxilla, the developing head induces bony formation where it initially lacked it. The maxilla continued its development, and no malocclusion developed as a result of cleft palate surgery repair.

The authors did not encounter any alveolar bone exostoses as reported in human dentistry [45,46].

Distalization of the maxillary canine tooth in one case in our study required the use of two OIs. Canine teeth are positioned in a transitional site on the dental arch, being located on the curve that reunites the lateral parts of the dental arch with the line of incisor teeth. The forces exerted on canine teeth are high, as noted by Zhou et al. [47]. Thus, their root morphology has adapted to high forces, resulting in noticeable volume. High orthodontic forces need to be applied to the canine tooth to be distalized as a result. This is the reason why the authors preferred to use two OIs to manage canine tooth distalization.

Regarding ethics in veterinary orthodontics, all dogs were neutered and removed from breeding if they were purebred [5,48]. The orthodontic treatment was carried out for functional reasons rather than cosmetic ones. Misaligned teeth can retain food, leading to increased dental plaque accumulation and a higher risk of periodontal diseases occurring earlier than in patients without dental issues. Moreover, conditions such as level and cross-bite are traumatic occlusions that can result in pathological attrition and complicated crown fractures over time, causing significant pain and discomfort [5,48]. The authors adhere to the ethical guidelines of the American Veterinary Medical Association (AVMA). The first principle of these guidelines emphasizes animal welfare as the top priority when using orthodontic treatments in small animals. The medical team was honest and transparent with the pet owners, presenting all potential complications that could arise during orthodontic treatment and ensuring informed consent. They clearly explained the potential benefits, justifying the associated risks and costs equitably [49]. The veterinary team implemented the best available procedures, which were imported from human medicine. It is widely recognized that any new medical procedures authorized for humans should first be thoroughly tested on animals, in alignment with the one health concept. Additionally, with a human dentist who is also a final-year veterinary medicine student serving as a consultant, the authors believe that the patients benefited from the most recent and validated medical techniques. The pain developed during orthodontic treatment with OIs was not noticeable and did not require any medication to be discontinued. The risk of nasal perforation in the third case was present, though a precise measurement ensured the cortical maintenance for OIs. Regarding the potential interference with craniofacial growth during the two-week maintenance of the custom-made palatal shield, the authors considered the risk to be minimal. Their hypothesis was later supported by the normal development of the head and the Class I relationship observed at the carnassials level on CT images taken two years after the surgery.

5. Conclusions

All cases involving the use of OIs had a successful outcome, without impacting dental and periodontal tissues. The good results can be attributed to strategic case selection and a more manageable number of cases, ensuring a higher quality of outcomes. The use of OIs may be a reliable solution for tooth movement in veterinary dentistry.

The innovative, custom-made PMMA shield, maintained with four dental OIs, may offer a reliable solution for the post-operative healing period when a surgical flap is exposed to the mechanical pressure applied by the tongue and its papillae.

Author Contributions: Conceptualization, R.-I.N., M.M.B., C.C.M. and I.M.; methodology, M.M.B.; software, R.-I.N. and M.M.B.; validation, R.-I.N., M.M.B., C.C.M. and I.M.; formal analysis, C.C.M.; investigation, M.M.B. and C.C.M.; resources, R.-I.N. and I.M.; data curation, I.M.; writing—original draft preparation, R.-I.N.; writing—review and editing, R.-I.N.; visualization, M.M.B., C.C.M. and I.M.; supervision, I.M.; project administration, I.M.; funding acquisition, R.-I.N. All authors have read and agreed to the published version of the manuscript.

Funding: This research received no external funding.

Institutional Review Board Statement: Decision of Ethical Board of University of Agricultural Studies and Veterinary Medicine Cluj-Napoca, no. 526/11.11.2025.

Informed Consent Statement: Informed written consent was obtained from all subjects involved in the study.

Data Availability Statement: The original contributions presented in this study are included in the article. Further inquiries can be directed to the corresponding author.

Conflicts of Interest: The authors declare no conflicts of interest.

Abbreviations

The following abbreviations are used in this manuscript:

OI	orthodontic implant
OIs	orthodontic implants
DI	dental implant
PS	primary stability
ISQ	implant stability quotation

References

1. Milella, L. Occlusion and malocclusion in the cat. *J. Feline Med. Surg.* **2014**, *17*, 5–20. [CrossRef] [PubMed]
2. Ziemann, D.; Mestrinho, L.A.; Gawor, J. Malocclusion in cats associated with mandibular soft tissue trauma: A retrospective case–control study. *J. Feline Med. Surg.* **2023**, *25*, 1098612X231202322. [CrossRef] [PubMed]
3. Thatcher, G. Diagnosis and management of Class II malocclusion. *Can. Vet. J.* **2019**, *60*, 791.
4. Baltayan, S.; Pi-Anfruns, J.; Aghaloo, T.; Moy, P.K. The predictive value of resonance frequency analysis measurements in the surgical placement and loading of endosseous implants. *J. Oral Maxillofac. Surg.* **2016**, *74*, 1145–1152. [CrossRef]
5. Reiter, A.M.; Gracis, M. *BSAVA Manual of Canine and Feline Dentistry and Oral Surgery*, 4th ed.; Cambrian Printers: Aberystwyth, UK, 2018.
6. Hershfeld, J.J.; Edward, H. Angle and the malocclusion of the teeth. *Bull. Hist. Dent.* **1979**, *27*, 79–84.

7. Devi, L.B.; Keisam, A.; Singh, H.P. Malocclusion and occlusal traits among dental and nursing students of Seven North-East states of India. *J. Oral Biol. Craniofac Res.* **2022**, *12*, 86–89. [CrossRef]
8. Masucci, C.; Oueiss, A.; Maniere-Ezvan, A.; Orthlieb, J.D.; Casazza, E. What is a malocclusion? *Orthod Fr.* **2020**, *91*, 57–67. [CrossRef]
9. Uzuner, F.D.; Aslan, B.I.; Dinçer, M. Dentoskeletal morphology in adults with Class I, Class II Division 1, or Class II Division 2 malocclusion with increased overbite. *Am. J. Orthod. Dentofac. Orthop.* **2019**, *156*, 248–256.e2. [CrossRef]
10. Ellis, E.; McNamara, J.A. Components of adult Class III malocclusion. *J. Oral Maxillofac. Surg.* **1984**, *42*, 295–305. [CrossRef]
11. AVDC. Available online: https://www.avdc.org (accessed on 30 October 2025).
12. Selvaraj, S.; Tandon, A.; Chandrasekaran, D.; Purushothaman, D.; Katepogu, P.; Mohan, R.; Angrish, N. Anchorage and Stability of Orthodontic Mini Implants in Relation to Length and Types of Implants. *Cureus* **2024**, *16*, e73056. [CrossRef]
13. Papadopoulos, M.A.; Tarawneh, F. The use of miniscrew implants for temporary skeletal anchorage in orthodontics: A comprehensive review. *Oral Surg. Oral Med. Oral Pathol. Oral Radiol. Endodontol* **2007**, *103*, e6–e15. [CrossRef] [PubMed]
14. Morais, L.S.; Serra, G.G.; Muller, C.A.; Andrade, L.R.; Palermo, E.F.; Elias, C.N.; Meyers, M. Titanium alloy mini-implants for orthodontic anchorage: Immediate loading and metal ion release. *Acta Biomater.* **2007**, *3*, 331–339. [CrossRef] [PubMed]
15. Available online: https://www.iso.org/obp/ui/#iso:std:iso:16443:ed-1:v1:en (accessed on 1 November 2025).
16. Elias, C.N.; Ruellas, O.; Fernandes, D.J. Orthodontic Implants: Concepts for the Orthodontic Practitioner. *Int. J. Dent.* **2012**, *2012*, 549761. [CrossRef] [PubMed]
17. Kanomi, R. Mini-implant for orthodontic anchorage. *J. Clin. Orthod.* **1997**, *31*, 763–767.
18. Silva, S.; Rayane, C.; Angelli, A.; Andrade, A.N.; Mendes-Marques, C.L.; Arruda, I.R.S.; Santos, L.R.L.; Vasconcelos, N.F.; Machado, G. Titanium Dental Implants: An Overview of Applied Nanobiotechnology to Improve Biocompatibility and Prevent Infections. *Materials* **2022**, *15*, 3150. [CrossRef]
19. Van den Berghe, F.; Cornillie, P.; Stegen, L.; Van Goethem, B.; Simoens, P. Palatoschisis in the dog: Developmental mechanisms and etiology. *Vlaams Diergeneeskd. Tijdschr.* **2010**, *79*, 117–124. [CrossRef]
20. Reiter, A.M.; Holt, D. Palate surgery. In *Veterinary Surgery: Small Animal*, 1st ed.; Tobias, K.M., Johnston, S.A., Eds.; Elsevier: St. Louis, MO, USA, 2012; pp. 1707–1717.
21. Beckman, B. Repair of the secondary cleft palate in the dog. *J. Vet. Dent.* **2011**, *28*, 58–62. [CrossRef]
22. Howard, D.R.; Davis, D.G.; Merkley, D.F.; Krahwinkel, D.J.; Schirmer, R.G.; Brinker, W.O. Mucoperiosteal flap technique for cleft palate repair in dogs. *J. Am. Vet. Med. Assoc.* **1974**, *165*, 352–354. [CrossRef]
23. Reiter, A.M.; Smith, M.M. The oral cavity and oropharynx. In *BSAVA Manual of Canine and Feline Head, Nech and Thoracic Surgery*, 1st ed.; Brockman, D.J., Holt, D.E., Eds.; BSAVA Publications: Gloucester, UK, 2005; pp. 25–43.
24. Khalil Ibrahim, M.; Hussein, A.A.; Mutlak, B.H. Morphological Features of Tongue in Domestic Cat Felis Catus. *Ibn AL-Haitham J. Pure Appl. Sci.* **2024**, *37*, 33–42. [CrossRef]
25. Liou, E.J.W.; Pai, B.C.J.; Lin, J.C.Y. Do miniscrews remain stationary under orthodontic forces? *Am. J. Orthod. Dentofac. Orthop.* **2004**, *126*, 42–47. [CrossRef]
26. Rokaya, D.; Srimaneepong, V.; Wisitrasameewon, W.; Humagain, M.; Thunyakitpisal, P. Peri-implantitis Update: Risk Indicators, Diagnosis, and Treatment. *Eur. J. Dent.* **2020**, *14*, 672. [CrossRef] [PubMed]
27. Bungău, T.C.; Moca, A.E.; Ciavoi, G.; Romanul, I.M.; Vaida, L.L.; Buhaș, C.L. Usage and Preferences of Orthodontic Mini-Implants Among Romanian Practitioners: A Survey Study. *Dent. J.* **2024**, *12*, 400. [CrossRef] [PubMed]
28. Tannenbaum, J.; Arzi, B.; Reiter, A.M.; Peralta, S.; Snyder, C.J.; Lommer, M.J.; Harvey, C.E.; Soukup, J.W.; Verstraete, F.J. The case against the use of dental implants in dogs and cats. *J. Am. Vet. Med. Assoc.* **2013**, *243*, 1680–1685. [CrossRef] [PubMed]
29. Available online: https://www.osstell.com/clinical-guidelines/the-osstell-isq-scale/ (accessed on 30 October 2025).
30. Sennerby, L.; Neil, M. Implant stability measurements using resonance frequency analysis: Biological and biomechanical aspects and clinical implications. *Periodontol.* **2000**, *47*, 51–66. [CrossRef]
31. Kim, J.W.; Ahn, S.J.; Chang, Y.I. Histomorphometric and mechanical analyses of the drill-free screw as orthodontic anchorage. *Am. J. Orthod. Dentofac. Orthop.* **2005**, *128*, 190–194. [CrossRef]
32. Gurdan, Z.; Szalma, J. Evaluation of the success and complication rates of self-drilling orthodontic mini-implants. *Niger. J. Clin. Pract.* **2018**, *21*, 546–552. [CrossRef]
33. Javali, M.A.; Betsy, J.; Safar, A.L.; Thobaiti, R.S.; Alshahrani, R.A.; AlQahtani, H.A.H. Relationship between Malocclusion and Periodontal Disease in Patients Seeking Orthodontic Treatment in Southwestern Saudi Arabia. *Saudi J. Med. Med. Sci.* **2020**, *8*, 133. [CrossRef]
34. Ingervall, B.; Jacobsson, U.; Nyman, S. A clinical study of the relationship between crowding of teeth, plaque and gingival condition. *J. Clin. Periodontol.* **1977**, *4*, 214–222. [CrossRef]
35. D'Cruz, E. Orthodontic treatment in periodontally compromised patients: Balancing risks, optimizing outcomes and guiding clinical practice. *MAG Online. Orthod. Update* **2025**, *18*, 138–144. [CrossRef]

36. Truong, V.M.; Kim, S.; Kim, J.; Lee, J.W.; Park, Y.S. Revisiting the Complications of Orthodontic Miniscrew. *BioMed Res. Int.* **2022**, *2022*, 8720412. [CrossRef]
37. Silveira, F.F.; Nunes, E.; Soares, J.A.; Ferreira, C.L.; Rotstein, I. Double 'pink tooth' associated with extensive internal root resorption after orthodontic treatment: A case report. *Dent. Traumatol. Off. Publ. Int. Assoc. Dent. Traumatol.* **2009**, *25*, e43–e47. [CrossRef]
38. Nedelea, R.-I.; Cezar Mătură, C.; Marcus, I. *Dental Implants In Cats and Dogs*; IntechOpen: London, UK, 2025. [CrossRef]
39. Brisceno, C.E.; Rossouw, P.E.; Carrillo, R.; Spears, R.; Buschang, P.H. Healing of the roots and surrounding structures after intentional damage with miniscrew implants. *Am. J. Orthod. Dentofac. Orthop.* **2009**, *135*, 292–301. [CrossRef] [PubMed]
40. Reynders, R.M.; Ladu, L.; Ronchi, L.; Di Girolamo, N.; de Lange, J.; Roberts, N.; Plüddemann, A. Insertion torque recordings for the diagnosis of contact between orthodontic mini-implants and dental roots: A systematic review. *Syst. Rev.* **2016**, *5*, 50. [CrossRef] [PubMed]
41. Kravitz, N.D.; Kusnoto, B. Risks and complications of orthodontic miniscrews. *Am. J. Orthod. Dentofac. Orthop.* **2007**, *131*, S43–S51. [CrossRef] [PubMed]
42. Motoyoshi, M.; Hirabayashi, M.; Uemura, M.; Shimizu, N. Recommended placement torque when tightening an orthodontic mini-implant. *Clin. Oral Implants Res.* **2006**, *17*, 109–114. [CrossRef]
43. Tan, A.; Heijdenrijk, K.; Moues, C.M. Custom-Made Palatal Shield Use in Cleft Palate and Fistula Repair: A Potential Benefit for Fast Postoperative Recovery. *Cleft Palate-Craniofacial J. Off. Publ. Am. Cleft Palate-Craniofacial Assoc.* **2018**, *55*, 307–311. [CrossRef]
44. Wright, A.L.; Peralta, S.; Fiani, N. Case report: Spontaneous mandibular body regeneration following unilateral subtotal mandibulectomy in a 3-month-old French bulldog. *Front. Vet. Sci.* **2023**, *10*, 1281232. [CrossRef]
45. Marquezan, M.; de Freitas, A.O.; Nojima, L.I. Miniscrew covering: An alternative to prevent traumatic lesions. *Am. J. Orthod. Dentofac. Orthop.* **2012**, *141*, 242–244. [CrossRef]
46. Manosudprasit, A.; Chantadilok, W.; Manosudprasit, M. Management of alveolar bone exostosis after orthodontic treatment. *J. Clin. Orthod.* **2021**, *55*, 401–413.
47. Zhou, Z.; Zhang, H.; Ren, Y.; Zhou, X.; Liu, H.; Zhong, Z.; Peng, G.; Cheng, J.; Zheng, C. Three-dimensional finite element analysis for best angles of dog canine teeth implantation. *Thai J. Vet. Med.* **2023**, *52*, 721–726.
48. Nedelea, R.I.; Borzan, M.M.; Dumitru, R.; Marcus, I. Two years of predictable design with computer-aided-design(CAD) and computer-aided manufacturing (CAM)orthodontic devices in rostral crossbites and levelbites in 12 dogs. *Rev. Romana Med. Veterinara* **2023**, *33*, 14–18.
49. American Veterinary Medical Association. Principles of Veterinary Ethics of the AVMA: The Principles with Supporting Annotations Va. 2015. Available online: https://avma.org/KB/Policies/Principles-of-Veterinary-Medical-Ethics-of-the-AMVA.aspx (accessed on 15 October 2025).

veterinary sciences

MDPI

Case Report

Effects and Microbiota Changes Following Oral Lyophilized Fecal Microbiota Transplantation Capsules in Canine with Chronic Enteropathy After Parvovirus Infection: Case Report

Siyu Liu [1], Baihui Zhou [2,3], Lei Liu [2,3], Jialai Zhong [4], Xinyan Zhang [5], Wenting Jiang [6], Haifeng Liu [1], Ziyao Zhou [1], Guangneng Peng [1], Yalin Zhong [1], Kun Zhang [1] and Zhijun Zhong [1,*]

1 Key Laboratory of Animal Disease and Human Health of Sichuan, College of Veterinary Medicine, Sichuan Agricultural University, Chengdu 611130, China; llsy6412024@163.com (S.L.); 410140017@163.com (H.L.); zzhou@sicau.edu.cn (Z.Z.); pgn.sicau@163.com (G.P.); 18190863559@163.com (Y.Z.); 15281297124@163.com (K.Z.)
2 Sichuan Anaerobic Biotechnology Co., Ltd., Chengdu 610200, China; zhoubaihui@scyysw.cn (B.Z.); liulei@scyysw.cn (L.L.)
3 Chengdu LifeMicro Technology Co., Ltd., Chengdu 610213, China
4 Biogas Institute of Ministry of Agriculture and Rural Affairs, Chengdu 610041, China; zhongjialai@163.com
5 College of Veterinary Medicine, Beijing University of Agriculture, Beijing 102206, China; xinyanzhang819@126.com
6 College of Landscape Architecture, Sichuan Agricultural University, Chengdu 611130, China; wentingj1010@163.com
* Correspondence: zhongzhijun488@126.com

Simple Summary: This case report described the effects of an oral, lyophilized capsule fecal microbiota transplantation (FMT) treatment on a dog suffering from chronic diarrhea following a canine parvovirus infection. Remarkably, the recipient's diarrhea resolved completely, and the stool consistency transitioned from a watery mucus-like state to a solid form post-FMT. Additionally, the recipient gained two kilograms in weight. We compared the fecal microbiomes of both the recipient and the donor, discussing our findings regarding bacterial engraftment in the recipient. Furthermore, an increase in metabolites (short-chain fatty acids) of the recipient was reported. Thus, oral lyophilized FMT capsules can be a promising treatment for dogs with chronic diarrhea that did not respond to conventional treatments.

Abstract: (1) Background: Chronic enteropathy (CE) in canines is associated with persistent microbiome dysbiosis, and conventional therapies (e.g., special diets, antimicrobials, and immunosuppressive drugs) are sometimes ineffective. Currently, fecal microbiota transplantation (FMT) has proven successful in treating CE in canines via invasive methods (e.g., enemas or endoscopy) or via oral frozen liquid capsules, which must be stored at −80 °C. However, due to the invasiveness of the administration methods and the storage constraints of the liquid capsules, FMT is not widely used in veterinary clinical practice. (2) Methods: The case of a four-year-old Siberian Husky with a three-year history of CE following canine parvovirus infection received lyophilized FMT capsules for thirty days. Stool samples were collected for metagenomic sequencing and quantification of fecal short-chain fatty acids (SCFAs), both pre- and post-FMT. Blood samples were analyzed using complete blood count (CBC) and biochemical testing. Ultrasound was used to assess the wall thickness of the stomach, duodenum, jejunum, and colon. (3) Results: Post-FMT, improvements in clinical outcomes were observed: fecal scores improved from 6 (unformed stools with mucus) to 2 (formed stool), and body weight increased by 8.3% (from 24.2 kg to 26.2 kg). Abnormal CBC and biochemical parameters were restored to reference ranges, including hematocrit (from 60.6% to 55.7%), hemoglobin (from 208 g/L to 190 g/L), creatinine (from

167 μmol/L to 121 μmol/L), and urea (from 11.9 mmol/L to 7.1 mmol/L). Ultrasound results showed that colonic wall thickness decreased from 0.23 ± 0.03 cm (pathological) to 0.18 ± 0.01 cm (physiological). Metagenomic analysis revealed that microbial richness (operational taxonomic units (OTUs) from 151 to 183) and diversity (Shannon and Simpson indices from 3.16 to 4.8 and from 0.87 to 0.94, respectively) all increased. The microbiota composition of the recipient exhibited a decline in the relative abundance of Firmicutes, falling from 99.84% to 35.62%, concomitant with an increase in Actinobacteria (from 0.08% to 4.78%), indicating a convergence toward a donor-like profile. Fecal SCFAs analysis revealed a 251.4% increase in propionate (from 0.0833 to 0.2929 mg/g) and elevated acetate (from 0.4425 to 0.4676 mg/g). These changes are functionally linked to enriched propanoate metabolism (Z = 0.89) in KEGG pathways. (4) Conclusions: Oral lyophilized FMT capsules resolved clinical signs of CE, enhanced microbial diversity and richness, and restored donor-like abundances of gut microbiota, particularly SCFA-producing taxa. Microbial restructuring increased microbial metabolite output, notably SCFA concentrations, and enriched functional metabolic pathways. Importantly, lyophilized FMT overcomes storage limitations and administration barriers, demonstrating its high clinical viability for treating canine CE.

Keywords: fecal microbiota transplantation; diarrhea; dogs; intestinal microbiome; short-chain fatty acids

1. Introduction

Chronic enteropathies (CE) are characterized by persistent gastrointestinal symptoms (e.g., vomiting and diarrhea) for ≥3 weeks, concurrent with intestinal dysbiosis [1]. Conventional CE therapies focus on specific diets (e.g., hydrolyzed protein diets) [2], antimicrobials (e.g., metronidazole or rifaximin) [3], and immunosuppressive drugs (e.g., prednisone) [4]. However, these can occasionally be ineffective, resulting in relapses [5,6]. Recent studies have shown that Fecal Microbiota Transplantation (FMT) can benefit patients by restoring intestinal microbial diversity and richness to a normal functional level [7–9]. FMT involves transferring the fecal microbiota of a healthy individual (donor) to a sick individual (recipient) to restore the recipient's gut microbiota [10]. FMT first attracted attention due to its efficacy in treating Clostridium difficile infection (CDI) in humans [11]. In veterinary clinical practice, some recent reports have demonstrated the efficacy of FMT against diseases such as inflammatory bowel disease (IBD), acute hemorrhagic diarrhea syndrome (AHDS), and CDI in dogs [12–15]. A previous study showed that FMT can also significantly accelerate recovery from diarrhea and reduce hospitalization time when used as an adjunctive therapy in the treatment of canine parvovirus (CPV) [16].

Multiple studies demonstrate that treating diarrheic dogs with FMT promotes microbial convergence toward donor profiles and enhances microbial diversity [17,18]. In dogs with IBD, which is characterized by *Firmicutes-* and *Proteobacteria*-dominated dysbiosis, FMT markedly increases the relative abundance of *Fusobacteriota* and drives microbial convergence toward donor-like profiles [19]. In AHDS, FMT accelerates the restoration of gut microbial diversity and significantly increases the abundance of bacteria (such as *Faecalibacterium* and *Fusobacterium*) that produce short-chain fatty acids (SCFAs) [20,21], which play a dominant role in intestinal barrier repair and have anti-inflammatory effects [13]. Therefore, the restoration of essential SCFA-producing bacteria during the process of FMT could benefit canines with CE.

Compared to the widespread use of FMT in humans, reports on the application of FMT in canine patients are limited. To date, only a few clinical veterinarians have used this technique [22,23]. This may be because FMT administration in dogs was initially carried out via endoscopy or enema, both of which are limited by their inconvenient and invasive procedures [24,25]. However, recent studies have further demonstrated the efficacy of oral frozen liquid suspension capsules in treating diarrheic canines, with comparable results to endoscopy and enema administration [14,26–28]. However, these capsules must be stored at −80 °C, which is inconvenient for pet owners. In this case, we evaluated the therapeutic efficacy of oral lyophilized FMT capsules in a canine with CE by longitudinal monitoring of clinic-based parameters (fecal score and body weight) and changes in gut microbiota pre- and post-FMT.

2. Clinical Case Description

Patient History and Clinical Presentation. A four-year-old, unneutered female Siberian Husky (named Hua Sheng) presented with chronic diarrhea and vomiting lasting three years and was diagnosed with CE at the Teaching Animal Hospital of Sichuan Agricultural University based on the established criterion of persistent gastrointestinal signs exceeding three weeks [1]. According to the owner's description, Hua Sheng was infected with canine parvovirus at one year of age. Following the resolution of the infection, the dog exhibited symptoms of chronic diarrhea, which have persisted to the present. Clinical manifestations included unformed stools with mucoid discharge and intermittent vomiting (two or three times per month). The following therapeutic interventions were administered: oral anthelmintic therapy (praziquantel), antimicrobial treatment (azithromycin), and systemic anti-inflammatory injections. However, only transient symptomatic improvement was observed, with no sustained resolution of clinical signs.

FMT Capsule Preparation. FMT capsules were prepared from fecal material donated by a 19-month-old male beagle (3.5 kg body weight), which was housed at an experimental breeding facility in Dujiangyan City, Sichuan Province. The facility implemented strict biosafety management practices involving daily disinfection protocols and restricted access. The donor was fed a standardized, commercially formulated diet that complied with the Chinese national standard GB 14924.1-2001 [29]. Prior to fecal sample collection, the donor underwent comprehensive health assessments: (1) complete blood count (CBC) and serum biochemistry profiles within the normal reference ranges; (2) an absence of parasites (Cryptosporidium, Giardia and Tritrichomonas foetus) in fecal examinations; and (3) no pathogens were detected in the screening for the three common diarrhea viruses: canine parvovirus, distemper virus and coronavirus.

The FMT capsules were manufactured in accordance with our previously established patent (CN118638635A) [30]. In brief, 100 g of fresh donor stool (wet weight) was homogenized with sterile physiological saline using a paddle homogenizer. The resulting suspension was then filtered sequentially through graded meshes (30 μm; 40 μm; 50 μm) to remove particulate matter. The filtrate was then centrifuged at $4000\times g$ for 10 min at 4 °C, and the resulting pellet was collected as the bacterial pellet. This pellet was blended with lyoprotectants (6% trehalose, 20% sucrose, 1% PEG-3350, 1% glycerol, 1% sodium glutamate, 2.5% sodium ascorbate, balance sterile water) at a mass ratio of 1:1.5 and emulsified to homogeneity. The mixture was then lyophilized for 48 h to produce microbial powder, which was pulverized and encapsulated in size 2 enteric-coated capsules (see Figure 1). The final products were stored at 2–8 °C. The total viable bacterial cell count in the FMT capsules is 4.11×10^8 CFU/g. Each capsule contained 100 mg of lyophilized powder, with a viable count of 4.11×10^7 CFU per capsule. The post-lyophilization survival rate exceeded 80%, and viability remained above 80% after 30 days of storage at 2–8 °C.

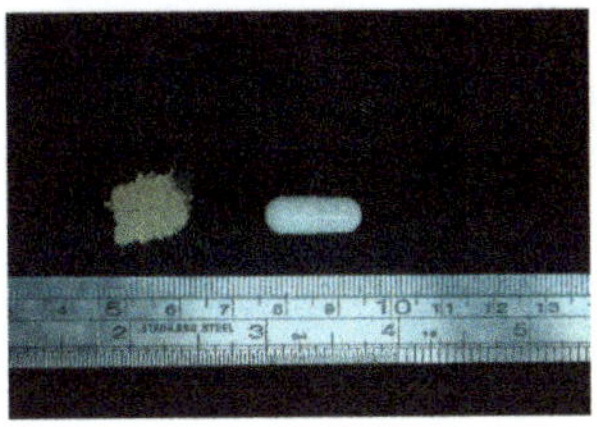

Figure 1. Oral lyophilized fecal microbiota transplantation capsules.

FMT Treatment Protocol. Prior to the present case, a pilot investigation by our team demonstrated that canines weighing kilograms or more who were given four capsules daily achieved better outcomes than those given two capsules daily. It is important to emphasize that this preliminary dosing strategy was exploratory and was not derived from an established or optimized protocol. Based on these findings, Hua Sheng was given four capsules orally each morning on an empty stomach for 30 days. CBC, biochemical analysis, C-reactive protein (CRP), and abdominal ultrasonography were tested at the Teaching Animal Hospital of Sichuan Agricultural University on Days 0, 15, and 30. Before FMT treatment, the recipient underwent a four-week drug-free washout period, which included all antimicrobials, anti-inflammatory drugs, and probiotics. During both the treatment and follow-up phases, the recipient was fed a consistent home-cooked diet (rabbit meat 290 g, potato 40 g, broccoli 25 g, oatmeal 25 g, bok choy 20 g, and flaxseed oil 2 g) and was prohibited from using any probiotics, antibiotics, or other gastrointestinal medications. CBC analysis was performed on venous blood samples (1 mL) and processed using an IDEXX ProCyte Dx analyzer to quantify erythrocyte, leukocyte, and platelet indices. Biochemical profiling was conducted on plasma obtained from heparinized whole blood (2 mL centrifuged at 4007× g for 3 min), with immediate quantification of alanine aminotransferase (ALT), alkaline phosphatase (ALP), blood urea nitrogen (BUN), and CRP levels via an IDEXX Catalyst One system. Ultrasonography was performed by a veterinary radiologist using a high-resolution system (GE Logiq E10 with an ML6-15-D linear probe at 15 MHz). Triplicate measurements were taken to assess the thickness of the gastric wall at the greater curvature (in a fasted state), the duodenal mucosa-submucosa complex in the proximal descending segment, and the jejunal and colonic walls within fluid-filled, non-peristaltic loops. According to the literature, the reference values for the duodenum, jejunum, and colon are 0.3–0.55 cm, 0.24–0.48 cm, and 0.11–0.19 cm, respectively [31]. Fecal samples collected during spontaneous defecation from dogs were cryopreserved at −80 °C for metagenomic and short-chain fatty acids (SCFAs) analysis. Daily fecal morphology was assessed using the validated Bristol Stool Form Scale (scale 1–7) [32].

FMT Efficacy Assessment via SCFAs Quantification and Metagenomic Profiling. The concentration of SCFAs was quantified using an Agilent 7890A gas chromatograph (Agilent Technologies, Santa Clara, CA, USA) according to the method described by Park HE et al. [33]. The stool samples were homogenized by transferring an accurate weight of 0.5 g of the sample to a 2 mL centrifuge tube. Add 1.2 mL of ultrapure water (the volume should be proportional to the sample mass), vortex to mix, then leave to stand for 5 min at room temperature. Centrifuge the suspension at 12,272× g for 15 min to precipitate insoluble debris. Then, 1 mL of the supernatant was mixed with 0.2 mL of 25% (w/v) metaphosphoric acid solution and 23.3 µL of 210 mmol/L crotonic acid solution (the volumes were adjusted proportionally to maintain the same ratio), incubated at 4 °C for 30 min, and then centrifuged at 4007× g for 10 min. Then, 0.3 mL of the clarified supernatant was diluted with 0.9 mL of chromatography-grade methanol (1:3 dilution ratio, adjusted to scale), vortexed, and centrifuged at 4007× g for 5 min. The final supernatant was filtered through a 0.22 μm nylon membrane and loaded into a gas chromatography vial for analysis. The conditions were as follows: HP-FFAP capillary column (30 m × 0.32 mm × 0.32 μm), using a high-purity nitrogen carrier gas at a flow rate of 2.2 mL/min. The inlet temperature was 250 °C. The oven programme started at 60 °C (held for 1 min), increased to 170 °C at a rate of 10 °C/min, and then to 212 °C at a rate of 8 °C/min (held for 2 min).

Following sample quality control (QC), 500 ng of metagenomic DNA was fragmented using the Covaris E220 system (Covaris, Brighton, UK). The fragmented DNA was then size-selected to a range of 300–700 bp using magnetic bead-based size selection. The selected DNA fragments underwent end repair and were ligated to indexed adapters. The ligation products were then amplified via PCR, hybridized with exon-specific probes, and captured using streptavidin-coated beads. The captured DNA underwent a second round of PCR amplification and was circularized to generate a single-stranded circular (ssCir) library. This ssCir library was then amplified using rolling circle amplification (RCA) to produce DNA nanoballs (DNBs). Finally, the DNBs were loaded onto a flow cell and sequenced using the DNBSEQ platform.

Metagenomic assembly and analysis. Metagenomic analyses began with a quality assessment of the raw sequencing data using FastQC (v0.12.1). Trimmomatic (v0.39) was then used for adapter and quality trimming with the parameters SLIDINGWINDOW:4:15 and MINLEN:99 to retain high-quality reads. Host-derived sequences were removed by aligning the reads to the reference genome of Canis lupus familiaris (GCF_011100685.1) using Bowtie2 (v2.5.4). The filtered reads were assembled de novo into contigs (≥500 bp) using MEGAHIT (v1.2.9) with iterative k-mer optimization (k-min: 27, k-max: 127). Open reading frames (ORFs) were predicted using Prodigal (v2.6.3) in metagenomic mode (-p meta). Functional annotation was performed against the KEGG database (KOfam release 2024) with stringent thresholds (E-value < 1×10^{-5} and coverage >70%). The ORFs were further annotated against the eggNOG 6.0 database (covering 5635 bacterial and archaeal taxa) using eggNOG-mapper v2.1.6 in Diamond mode to assign Clusters of Orthologous Groups (COG) categories. MetaPhlAn4 (v4.0.2) was used to quantify species-level relative abundance, and absolute abundance was derived by multiplying relative abundance by sample-specific, quality-controlled read counts. Alpha diversity was assessed using Pielou's evenness ($J = H'/\ln(S)$), Shannon entropy ($H = -\sum pi*\ln(pi)$), and Simpson index ($D = 1 - \sum(ni/N)^2$). Beta diversity was computed using QIIME 2 (v2024.2). Visualizations were generated using GraphPad Prism (v10.0) and TBtools (v2.0).

3. Results

3.1. Clinical Outcomes

3.1.1. Hematological Parameters

Prior to FMT, CBC analysis indicated that most hematological parameters were within the normal reference ranges (Table 1). Only four parameters were abnormal: hemoglobin (HGB, 208 g/L), hematocrit (HCT, 60.6%), mean corpuscular volume (MCV, 74.1 fL), and mean corpuscular hemoglobin (MCH, 25.4 pg). Post-FMT, all indicators were within the reference ranges, with all of the previously abnormal parameters (HGB, HCT, MCV, and MCH) returning to normal levels. Pre-FMT, biochemical profiling revealed that all biochemical profiling and CRP levels remained within the reference ranges, except for creatinine (CREA, 167.0 μmol/L) and urea (UREA, 11.9 mmol/L) (Table 2). Post-FMT, both CREA (121.0 μmol/L) and UREA (7.1 mmol/L) declined, falling within the reference ranges. Overall, FMT effectively restored all abnormal hematological and biochemical parameters to physiological levels.

Table 1. CBC values at baseline (T0) and post-FMT time points (T15 and T30) in the recipient.

Indicators	Unit	FMT-T0	FMT-T15	FMT-T30	Reference
HGB	g/L	208	192	190	110–190
HCT	%	60.6	55.1	55.7	39.0–56.0
WBC	10^9/L	11.1	8.4	10.5	6.0–17.0
Lymph	10^9/L	1.8	2.0	2.3	0.8–5.1
Mon	10^9/L	0.6	0.4	0.7	0.0–1.8
Gran	10^9/L	8.7	6.0	7.5	4.0–12.6
Lymph	%	16.4	23.9	21.6	12.0–30.0
Mon	%	5.2	5.3	6.8	2.0–9.0
Gran	%	78.4	70.8	71.6	60.0–83.0
RBC	10^{12}/L	8.18	7.53	7.58	5.50–8.50
MCV	fL	74.1	73.3	72.0	62.0–72.0
MCH	pg	25.4	25.4	25.0	20.0–25.0
MCHC	g/L	343	348	341	300–380
RDW	%	12.4	12.6	12.1	11.0–15.5
PLT		201	178	192	117–460
MPV	fL	8.3	8.6	8.3	7.0–12.9
PDW	10^9/L	16.6	17.0	16.6	
PCT	%	0.166	0.153	0.159	
Eos%	%	4.3	8.3	3.7	

Table 2. Biochemical parameters and CRP at baseline (T0) and post-FMT time points (T15 and T30) in the recipient.

Indicators	Unit	FMT-T0	FMT-T15	FMT-T30	Reference
GLU	mmol/L	5.82	5.75	6.06	4.11–7.95
CREA	μmol/L	167.0	119.0	121.0	44–159
UREA	mmol/L	11.9	6.8	7.1	2.5–9.6
BUN/CREA		18	14	14	
TP	g/L	68	66	66	52–82
ALB	g/L	35	33	33	23–40
GLOB	g/L	34	33	34	25–45
ALB/GLOB		1.0	1.0	1.0	
ALT	U/L	51	46	46	10–125
ALKP	U/L	44	46	50	23–212
CRP	mg/L	4.5	3.5	4.0	0.0–10.0

3.1.2. Fecal Score and Body Weight Recovery

As Figure 2 shows, pre-FMT, the fecal score was 6 (unformed feces with mucus). From T0 to T5 post-FMT, the fecal score decreased to 2.5. From T5 to T20 post-FMT, there was a fluctuating downward trend (scores 2–4), characterized by intermittent soft stools and minimal mucus. Between T20 and T30 post-FMT, fecal score stabilized at 2 (formed feces). Figure 3 shows the details of the change in body weight during FMT. From T0 to T5, body weight remained at 24.2 kg. From T5 to T15, there was a moderate increase in body weight (from 24.2 to 25.6 kg). By T30, the body weight had increased to 26.2 kg. Our results showed that the resolution of diarrhea occurred alongside sustained weight restoration.

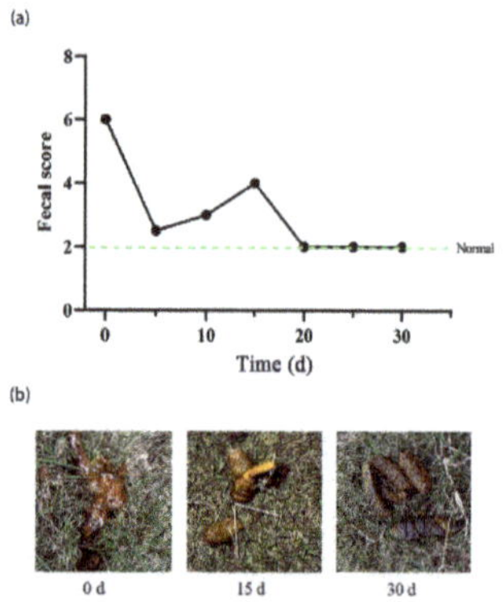

Figure 2. Restoration of fecal morphology following FMT therapy in the recipient dog. (**a**) Dynamic changes in fecal scores assessed by modified Bristol Stool Form Scale (range 1–7, dashed green line: physiological threshold at score 2). (**b**) Fecal samples at critical time points: T0 (unformed stools with mucus, score 6), T15 (semi-formed stools with mucus, score 4), and T30 (formed stools, score 2).

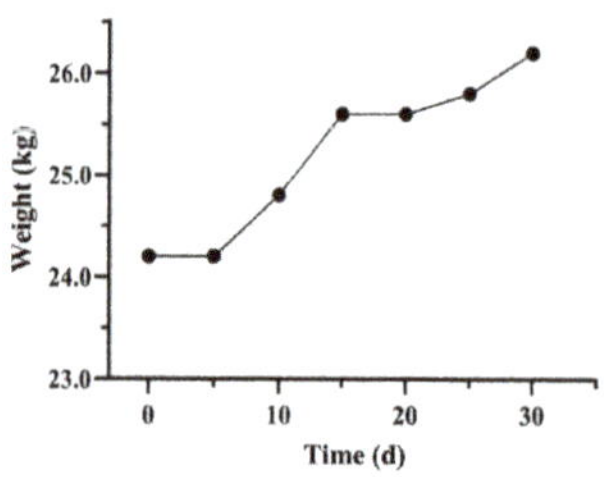

Figure 3. Body weight changes in the recipient treated with 30 days of FMT.

3.1.3. Gastrointestinal Morphology Evaluation

As Figure 4 shows, pre-FMT, the wall thicknesses of the stomach (0.33 ± 0.01 cm), duodenum (0.51 ± 0.02 cm), and jejunum (0.43 ± 0.03 cm) were all within the reference ranges (duodenum: 0.3–0.55 cm; jejunum: 0.24–0.48 cm). Only the colonic wall thickness (0.23 ± 0.03 cm) was outside the reference range (0.11–0.19 cm). Post-FMT, colonic wall thickness (0.16 ± 0.01 cm) returned to the normal range. Our results showed that FMT effectively restored pathological colonic wall thickness to standard physiological levels.

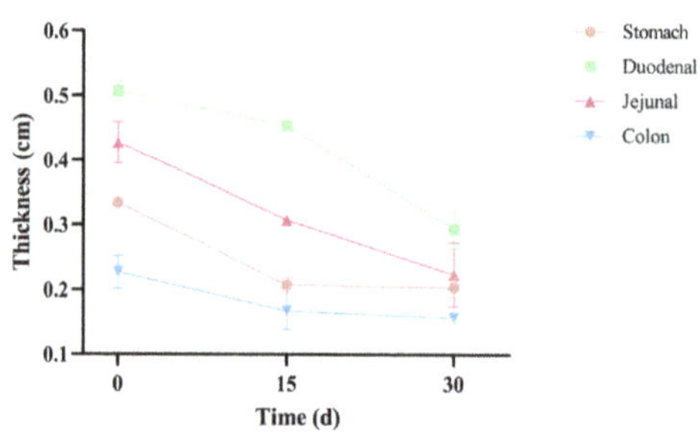

Figure 4. Ultrasonographic measurement of gastrointestinal wall thickness during FMT treatment. Line graphs depict changes in wall thickness (cm) of the stomach (pink), duodenum (green), jejunum (purple), and colon (blue) at baseline (T0) and T15 and T30 post-FMT administration in the recipient dog. Data points represent mean thickness ± SD (error bars).

3.2. Effects of FMT on Fecal Microbial Communities

3.2.1. Microbial Composition

Figure 5 shows the dynamic shifts in the recipient of gut microbiota during FMT at T0, T15, and T30. The relative abundances of all microbiota composition data are provided in Supplementary Table S1. At the phylum level (Figure 5a), the donor microbiota was dominated by *Firmicutes* (29.84%) and *Actinobacteria* (34.10%). By contrast, the recipient's microbiota pre-FMT exhibited extreme *Firmicutes* dominance (99.84%), which declined to 35.62% at T30 post-FMT. This was concurrent with an increase in *Actinobacteria* (from 0.08% to 4.78%). At the family level (Figure 5b), post-FMT, the relative abundances of *Erysipelotrichaceae* (from 0.06% to 4.96%), *Selenomonadaceae* (from 0.00% to 2.70%), *unclassified Firmicutes* (from 0.67% to 9.30%), *Coriobacteriaceae* (from 0.02% to 9.30%), and *Lachnospiraceae* (from 5.54% to 16.55%) increased, while the relative abundance of *Clostridiaceae* decreased from 93.22% to 0.23%, eventually approaching donor-like abundances (3.23%, 1.84%,1.87%, 33.69%, 15.44% and 1.87%, respectively). At the species level, the relative abundances of *Megamonas funiformis* (from 0% to 2.7%), *Allobaculum stercoricanis* (from 0% to 4.64%), *Blautia hansenii* (from 1.68% to 2.64%), *Ruminococcus gnavus* (from 0.06% to 5.31%), and *Blautia_SGB59878* (from 0.00% to 3.47%) increased and converged toward donor-like abundances (1.84%, 2.62%, 2.10%, 2.94%, and 2.47%, respectively). Notably, genus and species analysis (Figure 5c,d) revealed a taxonomic shift in the recipient's dominant microbiota: *Clostridium perfringens* was the most prevalent species pre-FMT (92.41%), while *Collinsella intestinalis* was the dominant species at T15 (42.98%), with *Escherichia coli* becoming dominant at T30 post-FMT (59.42%). Our results showed that the relative abundance of the recipient's microbiota post-FMT converged toward donor-like abundances.

Alterations in microbiome composition were further manifested by the overlap of operational taxonomic units (OTUs) between the donor and recipient. At the family level (Figure 6a), a total of 52 OTUs were observed. Ten OTUs were shared by all the samples, and the number of unique OTUs for the recipient at T15 (14) was higher than in other samples. At the genus level (Figure 6b), 14 OTUs were shared by all the samples. The number of unique OTUs for the donor at baseline and for the recipient at T0, T15, and T30 was 17, 1, 49, and 2, respectively. At the species level (Figure 6c), the number of shared OTUs between the recipient and the donor was just 14; most OTUs were unique, with 87 unique OTUs observed in the recipient at T15. Notably, the recipient at T15 exhibited maximal taxonomic overlap with the donor at family, genus, and species levels.

3.2.2. Diversity Analysis

As shown in Figure 7a,b, pre-FMT, the recipient's Shannon and Simpson indices (0.49 and 0.12, respectively) were lower than the donor's (4.18 and 0.87, respectively). At T15 post-FMT, both Shannon and Simpson indices demonstrated a peak in the recipient (3.32 and 0.75, respectively), approaching donor-like levels. Post-FMT, Shannon and Simpson indices showed a modest decline (2.17 and 0.56, respectively), yet remained elevated compared to pre-FMT levels. Figure 6c shows a similar trend in the observed species of the recipient during FMT. Pre-FMT, the recipient's microbiota exhibited low species richness (22 OTUs), surging to 167 OTUs at T15 and exceeding donor-like levels (109 OTUs). Post-FMT, richness declined to 44 OTUs, though this remained higher than pre-FMT levels. From T0 to T30, the Pielou index showed a fluctuating upward trend, from 0.11 to 0.40, approaching the donor-like level (0.62). These results demonstrate that FMT initially led to a surge in microbial diversity and richness, followed by partial stabilization, while achieving sustained improvement in community evenness.

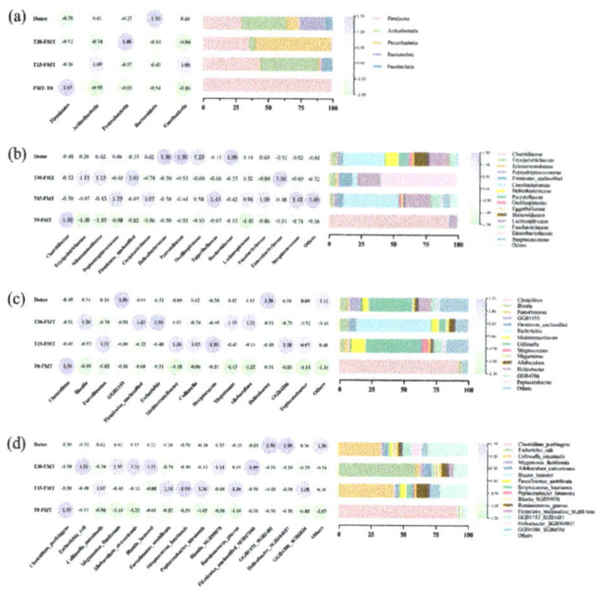

Figure 5. Analysis of intestinal microbial composition. (**a**) Left panel displays a heatmap of phylum-level abundances (graded blue to purple, indicating low-to-high relative abundance). Right panel shows a horizontal stacked bar chart of phylum composition. (**b**) Left panel displays a heatmap of family-level abundances. Right panel shows a horizontal stacked bar chart of the top fifteen most abundant taxa of family composition. (**c**) Left panel displays a heatmap of genus-level abundances. Right panel shows a horizontal stacked bar chart of the top fourteen most abundant taxa of genus composition. (**d**) Left panel displays a heatmap of species-level abundances. Right panel shows a horizontal stacked bar chart of the top fifteen most abundant taxa of species composition. All heatmaps and horizontal stacked bar charts include donor (baseline) and FMT timepoints (T10, T15, T30).

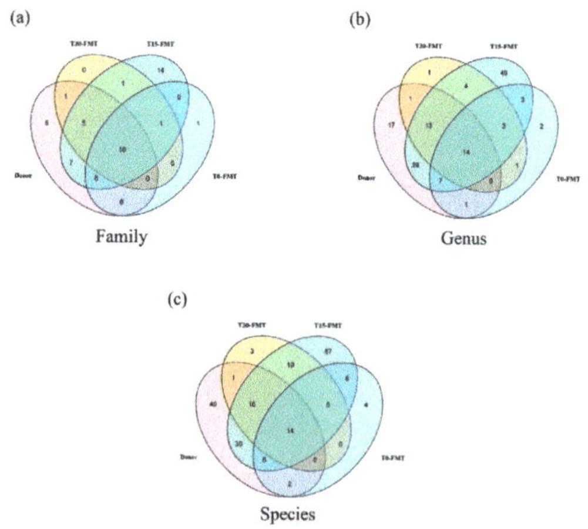

Figure 6. Venn diagrams showcasing the unique and shared microbiota among the donor at baseline and the recipient at T0, T15, and T30. In the Venn diagram, the number within the overlapping circles denotes the count of the shared OTUs. (**a**) Venn diagram on family level. (**b**) Venn diagram on genus level. (**c**) Venn diagram on species level.

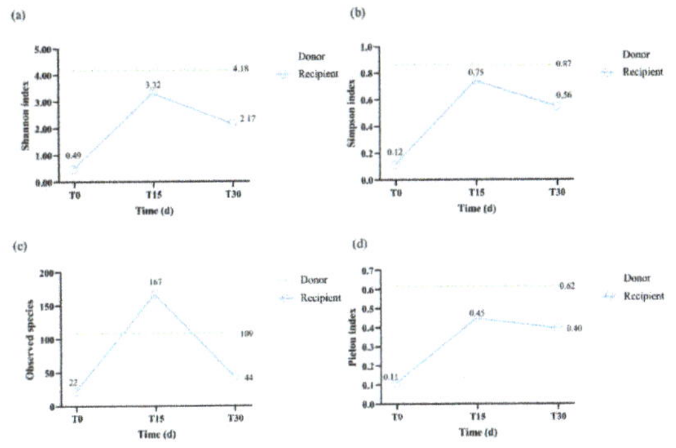

Figure 7. Alpha diversity analysis: (**a**) the values of the Shannon index; (**b**) the values of the Simpson index; (**c**) the values of the observed species; (**d**) the values of the Pielou index. The values were shown as a blue polyline for the recipient, and the reference values for the donor are shown in dashed green.

To further investigate differences in microbial structure between recipients and donors, beta diversity analysis was performed at T0, T15, and T30. Principal coordinate analysis (PCoA) showed that the first two principal coordinates (PC1 and PC2) together explained 92.50% of the beta diversity variance (Figure 8). Bray–Curtis distances showed that d_3 (0.77; donor vs. post-FMT recipient) and d_2 (0.40; donor vs. T15 post-FMT recipient) were smaller than d_1 (0.98; donor vs. pre-FMT recipient). These results suggest that FMT effectively restructured the composition of the recipient's gut microbiome, driving it toward a donor-like profile.

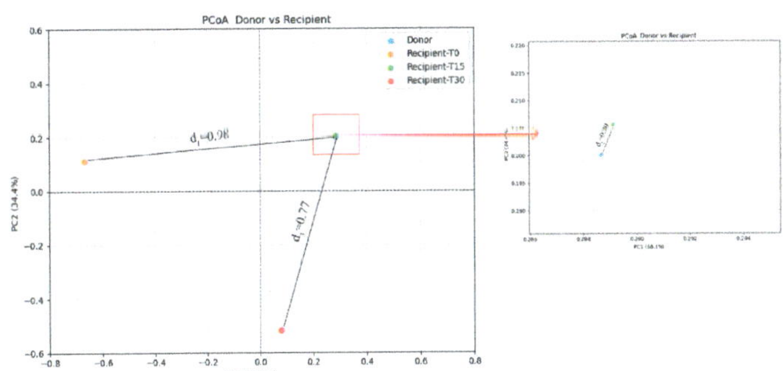

Figure 8. Principal coordinate analysis (PCoA) of the beta diversity of gut microbiota. The left panel shows the full-scale ordination of all samples (PC1 = 58.1%, PC2 = 34.4%; cumulative variance = 92.5%). Solid lines indicate the distances d1 (donor vs. pre-FMT recipient) and d3 (donor vs. T30 recipient). The right panel provides an enlarged view of the red boxed region, revealing clustering of the donor and recipient samples at T15. The distance d2 (donor vs. T15 recipient) is measured under magnification.

3.3. SCFAs and Metagenome Functions

To evaluate the impact of FMT on key SCFAs, we quantified the concentrations of acetate (AA), propionate (PA), and butyrate (BA) at T0 and T30. Pre-FMT, the concentrations of AA, PA, and BA were 0.4425 mg/g, 0.0833 mg/g, and 0.2971 mg/g, respectively. Post-FMT, AA and PA increased to 0.4676 mg/g (+5.7%) and 0.2929 mg/g (+251.4%), respectively, whereas BA decreased by 36% to 0.1900 mg/g.

To further investigate functional alterations in the metagenome associated with SCFAs metabolism, we performed Kyoto Encyclopedia of Genes and Genomes (KEGG) enrichment analysis and Cluster of Orthologous Groups of proteins (COG) functional analysis. As shown in Figure 9a, KEGG identified the top 35 significant pathways. The relative abundances of all KEGG pathways are provided in Supplementary Table S2. During FMT, direct pathways related to fatty acid metabolism showed significant upregulation, including propionate metabolism (z-score ranging from 0.20 to 0.89), pyruvate metabolism (z-score ranging from −0.84 to −0.26), and butanoate metabolism (z-score ranging from −1.14 to 0.44). Additionally, post-FMT, indirect pathways of fatty acid metabolism also showed upregulation, such as starch and sucrose metabolism (z-score ranging from −0.95 to 1.04) and glycolysis/glucogenesis (z-score ranging from −0.84 to −0.18). COG functional analysis (Figure 9b) revealed that the top 5 abundant pathways were L [Cell cycle control], K [Transcription], G [Carbohydrate transport and metabolism], M [Cell wall/membrane biogenesis], and C [Energy production and conversion]. Post-FMT, the dominant pathways shifted toward G, J [Translation, ribosomal structure and biogenesis], K, L, and M, with the relative abundances of pathways G, L, and M increasing, and those of C, J, and K declining. The relative abundances of all COG categories are provided in Supplementary Table S3.

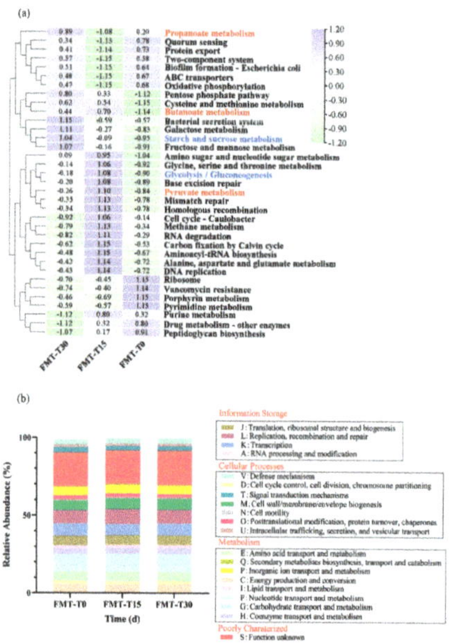

Figure 9. Functional prediction of gut microbiota via metagenomics. (**a**) Heatmap of KEGG pathways significantly altered post-FMT. Pathways directly and indirectly associated with fatty acid metabolism are highlighted in red and blue text, respectively. (**b**) Stacked bar plot of COG functional category abundances across timepoints.

4. Discussion

Many studies demonstrate that microbial dysbiosis consistently accompanies CE, subsequently inducing chronic diarrhea [34–36]. The 'CE-dysbiosis-diarrhea' cycle constitutes a self-perpetuating pathological loop. Restoring the gut microbiota is a vital intervention for breaking this cycle. A previous study reported that those who survive CPV infection have a significantly higher risk (odds ratio = 5.33) of developing chronic gastrointestinal disease [37]. In this case, we successfully treated a canine with CE that had persisted for three years following canine parvovirus infection using oral lyophilized FMT capsules, with no recurrence observed over a 6-month follow-up period (fecal scores consistently maintained at 2).

Fecal score and body weight are clinically relevant and visually intuitive indicators of diarrheal recovery [38]. In the present study, the fecal score decreased from 6 (unformed, mucus-covered stool) to 2 (formed, healthy stool), with no recurrence observed over a 6-month follow-up period. A previous study also showed that fecal scores normalized to 2 in dogs treated with FMT [15]. Body weight increased from 24.2 to 26.2 kg, and remained 26.3–26.8 kg over a 6-month follow-up period. Several studies have demonstrated that FMT enhances the body weight of patients [39,40]. Change in other laboratory indicators (hematological indices and ultrasonographic intestinal wall thickness) suggested that the FMT intervention achieved clinically relevant improvements.

The therapeutic benefit of FMT lies in its capacity to restore intestinal microbial diversity and richness, thereby re-establishing microecological balance in recipients [7,41,42]. Alpha and beta diversity were commonly used to reflect microbial diversity and richness. In our present study, the alpha diversity indices (Pielou, Shannon, and Simpson; see Figure 6a,b) revealed an increase in microbial diversity post-FMT (0.40, 2.17, and 0.56, respectively) compared to the pre-FMT values (0.11, 0.49, and 0.12). This suggests that FMT can reverse the reduction in gut microbiota richness induced by diarrhea. We observed that the Shannon and Simpson indices exhibited an initial increase followed by a subsequent decline during the FMT, which was consistent with a previous study [43]. Transient peaks in the Shannon and Simpson indices at T15 (see Figure 6b) correspond with the established engraftment dynamic whereby the initial introduction of donor microbiota leads to an increase in diversity, followed by selective retention of engrafted taxa [43,44]. Regarding beta diversity, PCoA results showed that the post-FMT microbiota were closer to the donor profile than the pre-FMT microbiota, indicating progressive convergence toward donor-like communities. Similarly, Goll et al. recently reported that FMT responders had a tendency to converge toward the donor microbiome profile after FMT, whereas this trend was less evident in non-responders [45]. This shift is consistent with established predictors of successful FMT, whereby increased alpha diversity and donor-like beta diversity are critical determinants of therapeutic efficacy [46]. The results of the beta diversity analysis were consistent with the observed changes in microbiota structure. In this case, the main shift occurred at the phylum level, reflecting a dynamic microbial restructuring process following FMT. Pre-FMT, *Firmicutes* (especially *C. perfringens*) was the most dominant phylum; at T15, *Actinobacteria* (especially *C. intestinalis*) became dominant; and post-FMT *Proteobacteria* (especially *E. coli*) became dominant at T30. This pattern of microbial succession may be driven by competitive exclusion, in which *C. perfringens* was initially replaced by *C. intestinalis* at T15, followed by *E. coli* at T30 [47,48]. Specifically, the expansion of E. coli at T30 occurred following a substantial decline in *C. intestinalis* (from 42.98% at T15 to 4.73% at T30), suggesting that competitive exclusion created a temporary niche opportunity that allowed *E. coli* to bloom. Thus, the transient dominance of E. coli can be interpreted as part of the ecological restructuring process, ultimately contributing to the restoration of microbial balance. These findings align with a previous study documenting the role of competitive exclusion in human CDI, whereby introduction of non-toxigenic strains reduces CDI recurrence [49]. Concurrently, FMT enriched beneficial microbial families (e.g., *Lachnospiraceae*, *Ruminococcaceae*, and *Erysipelotrichaceae*), which have the ability to produce SCFAs and induce anti-inflammatory Treg responses [50,51]. Specifically, the SCFA-producing genera *Blautia*, *Peptacetobacter*, and *Ruminococcus* are key microbial groups for canine intestinal health and play an important role in digesting complex carbohydrates, enhancing immunologic function, and preventing pathogenetic invasion [52,53]. In our case, the levels of *Blautia*, *Peptacetobacter*, and *Ruminococcus* microbial groups increased variably post-FMT and approached donor-like levels, suggesting that these microbial groups may have played a vital role in repairing the intestinal microbiota community and alleviating CE symptoms. In summary, FMT has been shown to disrupt the dysbiosis-inflammation cycle, thereby facilitating the restoration of microbial homeostasis. This restoration enhances microbial diversity and richness and enriches SCFA-producing microbes.

In addition to increasing the diversity and richness of the microbiota following FMT, bacterial metabolites such as SCFAs benefit the recipient by enhancing the intestinal barrier and energy metabolism [54,55]. We analyzed the three main SCFAs (acetate, butyrate, and propionate) at T0 and T30 to evaluate the restoration of metabolic activity in microbiota. Our results showed that the concentrations of two SCFAs (acetate and propionate) increased (from 0.4425 to 0.4676 mg/g and from 0.0833 to 0.2929 mg/g, respectively). Acetate and propionate positively affect glycolysis and propionate metabolism [56]. KEGG enrichment analysis revealed that glycolysis and propionate metabolism were significantly increased, suggesting that elevated acetate and propionate levels enhance carbohydrate metabolism and provide additional energy for the body [56]. A similar conclusion was obtained in a previous study, which found that propionate levels were increased significantly ($p = 0.025$) post-FMT [57]. Butyrate has been shown to possess anti-inflammatory properties, thereby promoting intestinal barrier function [58,59]. However, our case report revealed a 36% decrease in butyrate concentration (from 0.2971 mg/g to 0.1900 mg/g). A previous study demonstrated that the phylum Firmicutes (particularly the species *C. perfringens*) has the capacity to produce butyrate [60]. In our case, we found a decline in *C. perfringens* (from 92.41% to 0.03%) abundance. We therefore hypothesize that the decreased butyrate concentration may be related to the reduction of *C. perfringens* abundance. It is important to note that this is a correlative observation based on a single case, and a causative relationship cannot be established here. Although the concentration of butyrate declined, enrichment of the KEGG pathway in butyrate metabolism post-FMT indicates that the recipient's intestinal microbes may have acquired the capacity to synthesize butyrate. Thus, we speculate that the sharp reduction in *C. perfringens* may not be sufficient to rapidly restore butyric acid concentrations within the short term. Thus, the concentration of butyrate may require a longer period to manifest. This interpretation remains speculative and warrants further investigation in larger cases to validate the underlying mechanisms.

The limitations of this case report are as follows: (1) the limited sample size; larger case studies should be validated; (2) use of a single donor; multi-donor protocols should be prioritized to ensure broader applicability; (3) short follow-up period (30 days), extended monitoring is needed to evaluate long-term efficacy and safety. Nevertheless, the observed clinical recovery (no diarrhea recurrence at 6 months) and microbial restoration support the potential of oral lyophilized FMT capsules, despite these limitations.

5. Conclusions

In this case report, we successfully treated a canine with CE that had persisted for three years following canine parvovirus infection using oral lyophilized FMT capsules. Clinical improvements were observed, including the cessation of diarrhea with no recurrence over a 6-month follow-up period, as well as body weight gain. Concurrently, Microbial analysis revealed the enhanced diversity and richness of the gut microbiota following FMT, along with an increase in SCFA-producing bacteria and restoration of SCFAs-related metabolic pathways. This case provides a proof-of-concept example that oral lyophilized FMT may supply a therapeutic alternative for chronic diarrhea in dogs.

Supplementary Materials: The following supporting information can be downloaded at https://www.mdpi.com/article/10.3390/vetsci12090909/s1, Table S1. Relative abundance of microbial composition; Table S2: Relative abundance of KEGG pathways; Table S3: Relative abundance of COG categories.

Author Contributions: Z.Z. (Zhijun Zhong): Conceptualization, Methodology, Software. S.L., J.Z., X.Z. and W.J.: Data curation, Visualization, Writing-Original draft preparation. B.Z. and L.L.: Investigation. H.L., Z.Z. (Ziyao Zhou), Y.Z., K.Z. and G.P.: Software, Validation. S.L. and Z.Z. (Zhijun Zhong): Writing-Reviewing and Editing. All authors have read and agreed to the published version of the manuscript.

Funding: This work was supported by the Study on Key Technologies for Conservation of Wild Giant Panda Populations and Its Habitats within Giant Panda National Park System (CGF2024001) and the National Key Research and Development Program of China (2018YFD0500900, 2016YFD0501009).

Institutional Review Board Statement: This work was approved by the Sichuan Agricultural University Animal Ethical and Welfare Committee (ETHICS number: 20230268).

Informed Consent Statement: Informed consent was obtained from the owner of Hua Sheng.

Data Availability Statement: The data presented in this study are available in the article/ Supplementary Materials. Raw metagenomic sequencing data are available in the NCBI SRA under BioProject accession [PRJNA1298833]. Further inquiries can be directed to the corresponding authors.

Conflicts of Interest: Baihui Zhou and Lei Liu are employees (and owns stock) in Sichuan Anaerobic Biotechnology Co., Ltd. and Chengdu LifeMicro Technology Co., Ltd. The remaining authors have no conflicts of interest to declare.

References

1. Mehain, S.O.; Haines, J.M.; Lee, P.M. Platelet indices as biomarkers for characterization and determination of severity in canine chronic enteropathy. *Vet. J.* **2019**, *248*, 37–41.
2. Wang, S.; Martins, R.; Sullivan, M.C.; Friedman, E.S.; Misic, A.M.; El-Fahmawi, A.; De Martinis, E.C.P.; O'Brien, K.; Chen, Y.; Bradley, C.; et al. Diet-induced remission in chronic enteropathy is associated with altered microbial community structure and synthesis of secondary bile acids. *Microbiome* **2019**, *7*, 126. [CrossRef] [PubMed]
3. Menozzi, A.; Dall'Aglio, M.; Quintavalla, F.; Dallavalle, L.; Meucci, V.; Bertini, S. Rifaximin is an effective alternative to metronidazole for the treatment of chronic enteropathy in dogs: A randomised trial. *BMC Vet. Res.* **2016**, *12*, 217.
4. Cristóbal, J.I.; Duque, F.J.; Usón-Casaús, J.M.; Ruiz, P.; Nieto, E.L.; Pérez-Merino, E.M. Effects of Allogeneic Mesenchymal Stem Cell Transplantation in Dogs with Inflammatory Bowel Disease Treated with and without Corticosteroids. *Animals* **2021**, *11*, 2061. [CrossRef]
5. Li, X.; Gao, X.; Hu, H.; Xiao, Y.; Li, D.; Yu, G.; Yu, D.; Zhang, T.; Wang, Y. Clinical Efficacy and Microbiome Changes Following Fecal Microbiota Transplantation in Children with Recurrent Clostridium Difficile Infection. *Front. Microbiol.* **2018**, *9*, 2622. [CrossRef]
6. Makielski, K.; Cullen, J.; O'Connor, A.; Jergens, A.E. Narrative review of therapies for chronic enteropathies in dogs and cats. *J. Vet. Intern. Med.* **2019**, *33*, 11–22. [CrossRef]
7. Cui, B.; Li, P.; Xu, L.; Peng, Z.; Xiang, J.; He, Z.; Zhang, T.; Ji, G.; Nie, Y.; Wu, K.; et al. Step-up fecal microbiota transplantation (FMT) strategy. *Gut Microbes* **2016**, *7*, 323–328. [CrossRef]
8. Taur, Y.; Coyte, K.; Schluter, J.; Robilotti, E.; Figueroa, C.; Gjonbalaj, M.; Littmann, E.R.; Ling, L.; Miller, L.; Gyaltshen, Y.; et al. Reconstitution of the gut microbiota of antibiotic-treated patients by autologous fecal microbiota transplant. *Sci. Transl. Med.* **2018**, *10*, eaap9489. [CrossRef]
9. Mooyottu, S.; Muyyarikkandy, M.S.; Yousefi, F.; Li, G.; Sahin, O.; Burrough, E.; Scaria, J.; Sponseller, B.; Ramirez, A. Fecal microbiota transplantation modulates jejunal host-microbiota interface in weanling piglets. *Microbiome* **2025**, *13*, 45. [CrossRef] [PubMed]
10. Niederwerder, M.C.; Constance, L.A.; Rowland, R.R.R.; Abbas, W.; Fernando, S.C.; Potter, M.L.; Sheahan, M.A.; Burkey, T.E.; Hesse, R.A.; Cino-Ozuna, A.G. Fecal Microbiota Transplantation Is Associated with Reduced Morbidity and Mortality in Porcine Circovirus Associated Disease. *Front. Microbiol.* **2018**, *9*, 1631. [CrossRef] [PubMed]
11. Porcari, S.; Maida, M.; Bibbò, S.; McIlroy, J.; Ianiro, G.; Cammarota, G. Fecal Microbiota Transplantation as Emerging Treatment in European Countries 2.0. *Adv. Exp. Med. Biol.* **2024**, *1435*, 85–99.
12. Niina, A.; Kibe, R.; Suzuki, R.; Yuchi, Y.; Teshima, T.; Matsumoto, H.; Kataoka, Y.; Koyama, H. Fecal microbiota transplantation as a new treatment for canine inflammatory bowel disease. *Biosci. Microbiota Food Health* **2021**, *40*, 98–104. [CrossRef]
13. Gal, A.; Barko, P.C.; Biggs, P.J.; Gedye, K.R.; Midwinter, A.C.; Williams, D.A.; Burchell, R.K.; Pazzi, P. One dog's waste is another dog's wealth: A pilot study of fecal microbiota transplantation in dogs with acute hemorrhagic diarrhea syndrome. *PLoS ONE* **2021**, *16*, e0250344. [CrossRef]

14. Chaitman, J.; Ziese, A.L.; Pilla, R.; Minamoto, Y.; Blake, A.B.; Guard, B.C.; Isaiah, A.; Lidbury, J.A.; Steiner, J.M.; Unterer, S.; et al. Fecal Microbial and Metabolic Profiles in Dogs with Acute Diarrhea Receiving Either Fecal Microbiota Transplantation or Oral Metronidazole. *Front. Vet. Sci.* **2020**, *7*, 192. [CrossRef] [PubMed]
15. Sugita, K.; Yanuma, N.; Ohno, H.; Takahashi, K.; Kawano, K.; Morita, H.; Ohmori, K. Oral faecal microbiota transplantation for the treatment of Clostridium difficile-associated diarrhoea in a dog: A case report. *BMC Vet. Res.* **2019**, *15*, 11. [CrossRef]
16. Pereira, G.Q.; Gomes, L.A.; Santos, I.S.; Alfieri, A.F.; Weese, J.S.; Costa, M.C. Fecal microbiota transplantation in puppies with canine parvovirus infection. *J. Vet. Intern. Med.* **2018**, *32*, 707–711. [CrossRef]
17. Cantas, L.; Goll, R.; Fenton, C.G.; Paulssen, R.H.; Sørum, H. Impact of fecal microbiota transplantation in dogs. *Front. Vet. Sci.* **2025**, *12*, 1505226. [CrossRef]
18. Chaitman, J.; Gaschen, F. Fecal Microbiota Transplantation in Dogs. *Vet. Clin. N. Am. Small Anim. Pract.* **2021**, *51*, 219–233. [CrossRef]
19. Niina, A.; Kibe, R.; Suzuki, R.; Yuchi, Y.; Teshima, T.; Matsumoto, H.; Kataoka, Y.; Koyama, H. Improvement in Clinical Symptoms and Fecal Microbiome After Fecal Microbiota Transplantation in a Dog with Inflammatory Bowel Disease. *Vet. Med.* **2019**, *10*, 197–201. [CrossRef] [PubMed]
20. Nishiwaki, H.; Ito, M.; Hamaguchi, T.; Maeda, T.; Kashihara, K.; Tsuboi, Y.; Ueyama, J.; Yoshida, T.; Hanada, H.; Takeuchi, I.; et al. Short chain fatty acids-producing and mucin-degrading intestinal bacteria predict the progression of early Parkinson's disease. *NPJ Park. Dis.* **2022**, *8*, 65. [CrossRef] [PubMed]
21. Zhou, X.Y.; Li, M.; Li, X.; Long, X.; Zuo, X.L.; Hou, X.H.; Cong, Y.Z.; Li, Y.Q. Visceral hypersensitive rats share common dysbiosis features with irritable bowel syndrome patients. *World J. Gastroenterol.* **2016**, *22*, 5211–5227. [CrossRef]
22. Takáčová, M.; Bomba, A.; Tóthová, C.; Micháľová, A.; Turňa, H. Any Future for Faecal Microbiota Transplantation as a Novel Strategy for Gut Microbiota Modulation in Human and Veterinary Medicine? *Life* **2022**, *12*, 723. [CrossRef]
23. Salavati Schmitz, S. Observational Study of Small Animal Practitioners' Awareness, Clinical Practice and Experience with Fecal Microbiota Transplantation in Dogs. *Top. Companion Anim. Med.* **2022**, *47*, 100630. [CrossRef]
24. Marclay, M.; Dwyer, E.; Suchodolski, J.S.; Lidbury, J.A.; Steiner, J.M.; Gaschen, F.P. Recovery of Fecal Microbiome and Bile Acids in Healthy Dogs after Tylosin Administration with and without Fecal Microbiota Transplantation. *Vet. Sci.* **2022**, *9*, 324. [CrossRef] [PubMed]
25. Lee, M.A.; Questa, M.; Wanakumjorn, P.; Kol, A.; McLaughlin, B.; Weimer, B.C.; Buono, A.; Suchodolski, J.S.; Marsilio, S. Safety profile and effects on the peripheral immune response of fecal microbiota transplantation in clinically healthy dogs. *J. Vet. Intern. Med.* **2024**, *38*, 1425–1436. [CrossRef] [PubMed]
26. Cerquetella, M.; Marchegiani, A.; Rossi, G.; Trabalza-Marinucci, M.; Passamonti, F.; Isidori, M.; Rueca, F. Case Report: Oral Fecal Microbiota Transplantation in a Dog Suffering from Relapsing Chronic Diarrhea-Clinical Outcome and Follow-Up. *Front. Vet. Sci.* **2022**, *9*, 893342.
27. Hirsch, B.E.; Saraiya, N.; Poeth, K.; Schwartz, R.M.; Epstein, M.E.; Honig, G. Effectiveness of fecal-derived microbiota transfer using orally administered capsules for recurrent Clostridium difficile infection. *BMC Infect. Dis.* **2015**, *15*, 191. [CrossRef]
28. Rojas, C.A.; Entrolezo, Z.; Jarett, J.K.; Jospin, G.; Martin, A.; Ganz, H.H. Microbiome Responses to Oral Fecal Microbiota Transplantation in a Cohort of Domestic Dogs. *Vet. Sci.* **2024**, *11*, 42.
29. *GB/T 14924.1-2001*; Laboratory Animals-General Quality Standard for Formula Feeds. General Administration of Quality Supervision, Inspection and Quarantine of the People's Republic of China: Beijing, China, 2001.
30. Sichuan Anaerobic Biotechnology Limited Liability Company. A Lyophilized Protective Agent for Intestinal Flora and Its Application. Patent CN118638635A, 13 September 2024.
31. Gladwin, N.E.; Penninck, D.G.; Webster, C.R. Ultrasonographic evaluation of the thickness of the wall layers in the intestinal tract of dogs. *Am. J. Vet. Res.* **2014**, *75*, 349–353. [PubMed]
32. Chumpitazi, B.P.; Self, M.M.; Czyzewski, D.I.; Cejka, S.; Swank, P.R.; Shulman, R.J. Bristol Stool Form Scale reliability and agreement decreases when determining Rome III stool form designations. *Neurogastroenterol. Motil.* **2016**, *28*, 443–448. [CrossRef]
33. Park, H.E.; Kim, Y.J.; Do, K.H.; Kim, J.K.; Ham, J.S.; Lee, W.K. Effects of Queso Blanco Cheese Containing Bifidobacterium longum KACC 91563 on the Intestinal Microbiota and Short Chain Fatty Acid in Healthy Companion Dogs. *Korean J. Food Sci. Anim. Resour.* **2018**, *38*, 1261–1272. [CrossRef]
34. Ziese, A.L.; Suchodolski, J.S. Impact of Changes in Gastrointestinal Microbiota in Canine and Feline Digestive Diseases. *Vet. Clin. N. Am. Small Anim. Pract.* **2021**, *51*, 155–169. [CrossRef]
35. Schmid, S.M.; Tolbert, M.K. Harnessing the microbiome: Probiotics, antibiotics and their role in canine and feline gastrointestinal disease. *Vet. Rec.* **2024**, *195*, 13–25. [CrossRef]
36. Huang, Z.; Pan, Z.; Yang, R.; Bi, Y.; Xiong, X. The canine gastrointestinal microbiota: Early studies and research frontiers. *Gut Microbes* **2020**, *11*, 635–654. [CrossRef]
37. Kilian, E.; Suchodolski, J.S.; Hartmann, K.; Mueller, R.S.; Wess, G.; Unterer, S. Long-term effects of canine parvovirus infection in dogs. *PLoS ONE* **2018**, *13*, e0192198. [CrossRef]

38. Gómez-Gallego, C.; Junnila, J.; Männikkö, S.; Hämeenoja, P.; Valtonen, E.; Salminen, S.; Beasley, S. A canine-specific probiotic product in treating acute or intermittent diarrhea in dogs: A double-blind placebo-controlled efficacy study. *Vet. Microbiol.* **2016**, *197*, 122–128. [CrossRef]
39. Ren, J.; Li, Y.; Ni, H.; Zhang, Y.; Zhao, P.; Xiao, Q.; Hong, X.; Zhang, Z.; Yin, Y.; Li, X.; et al. Gut microbiota derived from fecal microbiota transplantation enhances body weight of Mimas squabs. *Anim. Biosci.* **2024**, *37*, 1428–1439. [CrossRef] [PubMed]
40. Lin, Z.L.; Lu, J.B.; Chen, Q.Y.; Cui, J.Q.; Ye, C.; Tian, H.L.; Qin, H.L.; Li, N. Clinical effectiveness of fecal microbiota transplantation combined with nutritional support and psychological intervention in patients with "Tetralogy of Tongji". *Zhonghua Wei Chang. Wai Ke Za Zhi* **2022**, *25*, 784–791.
41. Xue, L.; Deng, Z.; Luo, W.; He, X.; Chen, Y. Effect of Fecal Microbiota Transplantation on Non-Alcoholic Fatty Liver Disease: A Randomized Clinical Trial. *Front. Cell Infect. Microbiol.* **2022**, *12*, 759306. [CrossRef] [PubMed]
42. Staley, C.; Kelly, C.R.; Brandt, L.J.; Khoruts, A.; Sadowsky, M.J. Complete Microbiota Engraftment Is Not Essential for Recovery from Recurrent Clostridium difficile Infection following Fecal Microbiota Transplantation. *mBio* **2016**, *7*, e01965-16. [CrossRef] [PubMed]
43. Berlanda, M.; Innocente, G.; Simionati, B.; Di Camillo, B.; Facchin, S.; Giron, M.C.; Savarino, E.; Sebastiani, F.; Fiorio, F.; Patuzzi, I. Faecal Microbiome Transplantation as a Solution to Chronic Enteropathies in Dogs: A Case Study of Beneficial Microbial Evolution. *Animals* **2021**, *11*, 1433. [CrossRef]
44. Weingarden, A.; González, A.; Vázquez-Baeza, Y.; Weiss, S.; Humphry, G.; Berg-Lyons, D.; Knights, D.; Unno, T.; Bobr, A.; Kang, J.; et al. Dynamic changes in short- and long-term bacterial composition following fecal microbiota transplantation for recurrent Clostridium difficile infection. *Microbiome* **2015**, *3*, 10. [CrossRef] [PubMed]
45. Goll, R.; Johnsen, P.H.; Hjerde, E.; Diab, J.; Valle, P.C.; Hilpusch, F.; Cavanagh, J.P. Effects of fecal microbiota transplantation in subjects with irritable bowel syndrome are mirrored by changes in gut microbiome. *Gut Microbes* **2020**, *12*, 1794263. [CrossRef] [PubMed]
46. Guo, Q.; Lin, H.; Chen, P.; Tan, S.; Wen, Z.; Lin, L.; He, J.; Wen, J.; Lu, S. Dynamic changes of intestinal flora in patients with irritable bowel syndrome combined with anxiety and depression after oral administration of enterobacteria capsules. *Bioengineered* **2021**, *12*, 11885–11897. [CrossRef] [PubMed]
47. Kelly, C.R.; Khoruts, A.; Staley, C.; Sadowsky, M.J.; Abd, M.; Alani, M.; Bakow, B.; Curran, P.; McKenney, J.; Tisch, A.; et al. Effect of Fecal Microbiota Transplantation on Recurrence in Multiply Recurrent Clostridium difficile Infection: A Randomized Trial. *Ann. Intern. Med.* **2016**, *165*, 609–616. [CrossRef]
48. Tuniyazi, M.; Hu, X.; Fu, Y.; Zhang, N. Canine Fecal Microbiota Transplantation: Current Application and Possible Mechanisms. *Vet. Sci.* **2022**, *9*, 396. [CrossRef]
49. Mullish, B.H.; Merrick, B.; Quraishi, M.N.; Bak, A.; Green, C.A.; Moore, D.J.; Porter, R.J.; Elumogo, N.T.; Segal, J.P.; Sharma, N.; et al. The use of faecal microbiota transplant as treatment for recurrent or refractory Clostridioides difficile infection and other potential indications: Second edition of joint British Society of Gastroenterology (BSG) and Healthcare Infection Society (HIS) guidelines. *Gut* **2024**, *73*, 1052–1075. [CrossRef]
50. Louis, P.; Flint, H.J. Formation of propionate and butyrate by the human colonic microbiota. *Environ. Microbiol.* **2017**, *19*, 29–41. [CrossRef]
51. Godínez-Méndez, L.A.; Gurrola-Díaz, C.M.; Zepeda-Nuño, J.S.; Vega-Magaña, N.; Lopez-Roa, R.I.; Íñiguez-Gutiérrez, L.; García-López, P.M.; Fafutis-Morris, M.; Delgado-Rizo, V. In Vivo Healthy Benefits of Galacto-Oligosaccharides from Lupinus albus (LA-GOS) in Butyrate Production through Intestinal Microbiota. *Biomolecules* **2021**, *11*, 1658. [CrossRef]
52. Wang, W.; Zheng, S.; Li, L.; Yang, Y.; Liu, Y.; Wang, A.; Sharshov, K.; Li, Y. Comparative metagenomics of the gut microbiota in wild greylag geese (*Anser anser*) and ruddy shelducks (*Tadorna ferruginea*). *Microbiologyopen* **2019**, *8*, e00725. [CrossRef]
53. Sarmiento, J.; Pulgar, R.; Mandakovic, D.; Porras, O.; Flores, C.A.; Luco, D.; Trujillo, C.A.; Díaz-Esquivel, B.; Alvarez, C.; Acevedo, A.; et al. Nocturnal Light Pollution Induces Weight Gain in Mice and Reshapes the Structure, Functions, and Interactions of Their Colonic Microbiota. *Int. J. Mol. Sci.* **2022**, *2*, 1673. [CrossRef]
54. Liu, P.; Wang, Y.; Yang, G.; Zhang, Q.; Meng, L.; Xin, Y.; Jiang, X. The role of short-chain fatty acids in intestinal barrier function, inflammation, oxidative stress, and colonic carcinogenesis. *Pharmacol. Res.* **2021**, *165*, 105420. [CrossRef] [PubMed]
55. Fan, P.; Nelson, C.D.; Driver, J.D.; Elzo, M.A.; Peñagaricano, F.; Jeong, K.C. Host genetics exerts lifelong effects upon hindgut microbiota and its association with bovine growth and immunity. *ISME J.* **2021**, *15*, 2306–2321. [CrossRef]
56. Yang, X.; Tai, Y.; Ma, Y.; Xu, Z.; Hao, J.; Han, D.; Li, J.; Deng, X. Cecum microbiome and metabolism characteristics of Silky Fowl and White Leghorn chicken in late laying stages. *Front. Microbiol.* **2022**, *13*, 984654. [CrossRef]
57. Osaki, H.; Jodai, Y.; Koyama, K.; Omori, T.; Horiguchi, N.; Kamano, T.; Funasaka, K.; Nagasaka, M.; Nakagawa, Y.; Shibata, T.; et al. Clinical response and changes in the fecal microbiota and metabolite levels after fecal microbiota transplantation in patients with inflammatory bowel disease and recurrent Clostridioides difficile infection. *Fujita Med. J.* **2021**, *7*, 87–98.
58. Donohoe, D.R.; Garge, N.; Zhang, X.; Sun, W.; O'Connell, T.M.; Bunger, M.K.; Bultman, S.J. The microbiome and butyrate regulate energy metabolism and autophagy in the mammalian colon. *Cell Metab.* **2011**, *13*, 517–526. [CrossRef]

59. Hamer, H.M.; Jonkers, D.; Venema, K.; Vanhoutvin, S.; Troost, F.J.; Brummer, R.J. Review article: The role of butyrate on colonic function. *Aliment. Pharmacol. Ther.* **2008**, *27*, 104–119. [CrossRef] [PubMed]
60. Vital, M.; Gao, J.; Rizzo, M.; Harrison, T.; Tiedje, J.M. Diet is a major factor governing the fecal butyrate-producing community structure across Mammalia, Aves and Reptilia. *ISME J.* **2015**, *9*, 832–843. [CrossRef] [PubMed]

veterinary sciences

MDPI

Article

Effects of Canthoplasty in Preventing Secondary Ocular Surface Lesions in Brachycephalic Dogs

Alcyjara Rêgo Costa [1], Tatiane Avelar Ribeiro [1], Diego Marques C. Silva [1], José Ribamar da Silva Júnior [2], Tiago Barbalho Lima [2], Rui M. Gil da Costa [3,4], Ana I. Faustino-Rocha [4,5,*], Paula A. Oliveira [4,6] and Ana Lúcia Abreu-Silva [3,*]

1 Veterinary Hospital, Universidade Estadual do Maranhão, 1.000, São Luís 65055-310, MA, Brazil; alcyjara.costa@hotmail.com (A.R.C.); tatianeavelar63422@gmail.com (T.A.R.); diegomarquescs@gmail.com (D.M.C.S.)
2 Departamento das Clínicas, Universidade Estadual do Maranhão, 1.000, São Luís 65055-310, MA, Brazil; aneju@gmail.com (J.R.d.S.J.); barbalho.tiago@gmail.com (T.B.L.)
3 Departamento de Patologia, Universidade Estadual do Maranhão, 1.000, São Luís 65055-310, MA, Brazil; ruioliveira@cca.uema.br
4 Centre for the Research and Technology of Agro-Environmental and Biological Sciences (CITAB), Institute for Innovation, Capacity Building and Sustainability of Agri-Food Production (Inov4Agro), University of Trás-os-Montes and Alto Douro (UTAD), 5000-801 Vila Real, Portugal; pamo@utad.pt
5 Department of Zootechnics, Comprehensive Health Research Center (CHRC), University of Évora, 7004-516 Évora, Portugal
6 Department of Veterinary Sciences, University of Trás-os-Montes and Alto Douro, 5000-801 Vila Real, Portugal
* Correspondence: anafaustino@uevora.pt (A.I.F.-R.); abreusilva.ana@gmail.com (A.L.A.-S.)

Simple Summary: When detected early, canthoplasty controls the clinical signs and complications associated with accentuated eye exposure. This study evaluated the reduction in the palpebral fissure to prevent ophthalmic lesions secondary to brachycephalic dog syndrome. A total of 64 eyes of brachycephalic dogs were studied, divided into the control group (22 eyes submitted to traditional clinical treatment) and the treated group (42 eyes submitted to canthoplasty). The treated group showed an improvement in tear production compared to the control group. In addition, the animals treated showed a significant increase in TFBUT (T9M), which was higher than the control group. The TCT evaluation scores of the treated group decreased compared to the control group (T12M). The occurrence of corneal ulcers was significantly higher in the control group in the medium term. We concluded that early canthoplasty in brachycephalic dogs prevented the occurrence of secondary lesions, increased tear production and quality, and reduced the occurrence of ulcers, particularly in the medium term.

Abstract: When detected early, canthoplasty controls the clinical signs and complications associated with accentuated eye exposure. The aim of this study was to evaluate the reduction in the palpebral fissure to prevent ophthalmic lesions secondary to brachycephalic dog syndrome. A total of 64 eyes of brachycephalic dogs were studied, divided into the control group (22 eyes submitted to traditional clinical treatment) and the treated group (42 eyes submitted to canthoplasty). Tear production, tear film break-up time (TFBUT), tear crystallization time (TCT), and the occurrence of corneal ulcers were assessed prior to treatment (T0), seven days after therapy or surgery (T7), and after 30 days (T30), and reassessments were performed every three months, until month 18. It was observed that the treated animals (treated group) showed an improvement in tear production compared to the control animals (control group). In addition, the treated animals showed a significant increase in TFBUT (T9M), which was higher than the control group. The TCT evaluation scores of the treated group decreased compared to the control group (T12M), and the occurrence of corneal ulcers was significantly higher in the control group in the medium term.

 https://doi.org/10.3390/vetsci12090889

We concluded that early canthoplasty in brachycephalic dogs prevented the occurrence of secondary lesions, increased tear production and quality, and reduced the occurrence of ulcers, particularly in the medium term.

Keywords: brachycephalic eye syndrome; canthoplasty; dog; ophthalmic surgery

1. Introduction

The artificial selection of dog breeds has led to the emergence of various disorders, including ophthalmic disorders, especially in brachycephalic dogs [1]. These breeds have morphological characteristics that predispose them to lesions on the ocular surface, such as macro-blepharon, euriblepharon, shallow orbits, short muzzles with skin folds, medial entropion, hair on the caruncle, and abnormal variations [2]. These factors increase the exposure of the eyeball and make it difficult for the eyeball to close (lagophthalmos), increasing the risk of pigmentary keratitis, keratoconjunctivitis sicca, proptosis, and corneal ulcers [3]. With the growing popularity of these breeds, veterinarians and regulatory bodies are seeking to reduce these characteristics and minimize the associated disorders. Surgical reduction in the palpebral fissure is indicated for breeds such as the Lhasa Apso, Pekingese and Shih Tzu, which have exophthalmos and impaired blink reflex, leading to corneal ulcers and changes in ocular lubrication. Medial, lateral, or combined canthoplasty reduces ocular exposure, improving control of complications [4]. Many canthoplasties are performed late when lesions are already severe and vision is compromised [5]. The study's hypothesis is that early canthoplasty may prevent damage to the ocular surface and prevent complications like dry keratoconjunctivitis and corneal ulcers. Considering the complications associated with the morphologic traits of brachycephalic breeds, combining surgical techniques to reduce the ocular surface is likely to help prevent such lesions in these canine breeds. The present study aims to assess the effects of canthoplasty with reduction in the palpebral fissure on the prevention of secondary ophthalmic lesions such as dry keratoconjunctivitis and corneal ulcers in brachycephalic dogs, and compare the results with standard clinical therapy.

2. Materials and Methods

The research was approved by the ethics committee (CEEA, protocol 35/2018) and carried out with the permission of the owners. Sixty-four eyes of brachycephalic dogs (6–24 months) treated at the Universidade Estadual do Maranhão (UEMA) Veterinary Hospital were analyzed. The ophthalmic examination included Schirmer's tear test (Madhu Innovating Possibilities, New Dehli, India), a tear crystallization test (TCT) (Perfecta, São Paulo, Brazil), biomicroscopy using a portable slim lamp (Shanghai Bolan Optical Electric, Shanghai, China), tonometry using a FA-800 tonometer (Electric Suoer, St. Louis, MO, USA), ophthalmoscopy (PanOptic Plus 11820, Welch Allyn, St. Louis, MO, USA), fluorescein (Madhu Innovating Possibilities, India), tear film break-up time (TFBUT), and measurement of the palpebral fissure with a caliper (Universal 150 mm caliper, Fort G, São Paulo, Brazil). Patients were selected based on morphological characteristics such as macroblepharon and lagophthalmos, excluding those with distichiasis, ectopic eyelashes, and eyelid tumors. Physical and laboratory examinations assessed systemic conditions.

The dogs were divided into a control group, which underwent clinical treatment, and a treatment group, which underwent medial/lateral canthoplasty and entropion correction. All received topical therapy with cyclosporine, lacrimomimetics, and anti-inflammatories.

After fasting for 12 h on food and six hours on water, the treated patients (treated group) were pre-anesthetized with meperidine (0.005 mg/kg, intramuscular) and diazepam (0.3 mg/kg). Meloxicam (0.1 mg/kg, subcutaneous) was administered as a single dose. After 15 min, anesthesia was induced with propofol (5 mg/kg, intravenous) and maintained with isoflurane in a semi-closed circuit.

The surgical procedure consisted of simple medial canthoplasty [6]. The patients were positioned in the sternal decubitus position and antisepsis was performed with iodopyrrolidone in aqueous solution and diluted (1:50) in Ringer's lactate. The upper and lower lacrimal canaliculi were identified and cannulated with a no. 24 catheter.

A triangular section of tissue around the lateral canthus medial was carefully removed, with the base oriented towards the medial commissure and the apex directed toward the lateral commissure. The base usually measures between 2 and 4 mm. Using a no. 15 scalpel or iris scissors, the marked triangle was carefully resected, removing the skin and conjunctiva. The incision was made at a distance of 1 to 2 mm from the lacrimal canaliculi, in order to preserve those structures. To approximate and align the edges, a figure-of-eight suture was made with 5-0 nylon. The skin was then sutured in a simple interrupted pattern with 5-0 nylon, as depicted in Figure 1.

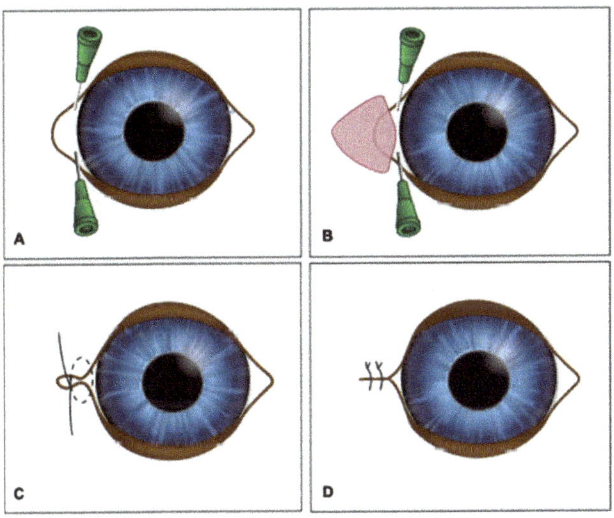

Figure 1. The medial canthoplasty technique. (**A**) Identification of the upper and lower lacrimal canaliculi; (**B**) resection of a skin and conjunctiva triangle, with an incision at 1–2 mm from the lacrimal canaliculi; (**C**) figure-of-eight suture to approximate and align the edges; (**D**) skin suture in a simple interrupted pattern.

Postoperatively, the patients received meloxicam (0.1 mg/kg SID, 4 days) and tramadol (4 mg/kg, 5 days). The surgical wound was cleaned with saline solution and healing ointment, and the use of an Elizabethan collar was maintained until the stitches were removed. Evaluations took place before treatment (T0), at 7 (T7) and 30 days (T30), and every 3 months, up to 18 months (T3M-T18M), in both groups (control and treated).

Tear production was assessed by the Schirmer test, performed before any ocular manipulation, using sterile filter paper strips inserted into the conjunctival sac for one minute. The tear film break-up test (TFBUT) was carried out by instilling fluorescein into the conjunctiva, followed by manual closure of the eyelids for 10 s. The time until the first tear film rupture was timed under illumination with a cobalt blue filter, and the measurement was repeated. For the tear crystallization test (TCT), samples were collected with a microcapillary tube from the lower conjunctival sac of both eyes. The contents were placed on glass slides and analyzed under optical microscopy at 4× and 10× magnification. The classification followed the proposal by Rolando [7], categorizing the findings into four groups according to the pattern of arborization and the distance between the crystals. The presence of corneal ulcers was compared between the groups using slit lamp biomicroscopy and the fluorescein test.

The parametric data was organized in a factorial scheme (2 × 9) and submitted to analysis of variance (ANOVA), with comparison of means using the Tukey test. Qualitative variables were analyzed using the Wilcoxon test (for comparisons between groups) and the Kruskal–Wallis test (for variation over time), adopting a significance level of $p < 0.05$.

3. Results

A total of 64 eyes from 32 male and female brachycephalic dogs aged between 6 months and 24 months (two years) were evaluated. Of this total, 43 eyes (22 animals) that had undergone the canthoplasty procedure, the treated group, and 22 eyes (11 animals) that had undergone clinical treatment, the control group, were evaluated. One animal in the treated group had previously undergone unilateral enucleation. One animal from the treated group dropped off from T7 onwards. At T6M, one control animal suffered severe corneal perforation and underwent unilateral enucleation. The n for treated and control groups were reduced, accordingly, to 42 eyes (21 animals) and 21 eyes (11 animals), respectively.

The animals evaluated were puppies or youngsters and were already showing at least one alteration in the ocular surface associated with ocular brachycephalic syndrome, albeit discreetly.

In the evaluation of tear production, a significant increase in the Schirmer test was reported seven days after surgery in the treated group, possibly due to the surgical manipulation. After 30 days (T30), there was a reduction, followed by a progressive increase in the medium term (Table 1).

Table 1. Schirmer's lacrimal test performed on brachycephalic animals submitted to clinical treatment (control group) or early blepharoplasty (treated group) throughout the evaluation periods. Data are presented as mean ± standard deviation (S.D.).

Groups	Time									Mean ± S.D. Overtime
	T0	T7	T30	T3M	T6M	T9M	T12M	T15M	T18M	
Control group	14.36 ± 7.91 ^{aA}	14.72 ± 7.72 ^{aA}	16.6 ± 5.88 ^{aA}	15.14 ± 5.47 ^{aA}	14.68 ± 5.65 ^{aA}	14.41 ± 4.67 ^{aA}	11.00 ± 4.90 ^{aA}	11.14 ± 4.83 ^{aA}	11.43 ± 4.89 ^{aA}	14.07 ± 6.23 ^{A}
Treated group	15.74 ± 8.85 ^{aA}	20.66 ± 7.05 ^{bB}	17.98 ± 7.09 ^{abA}	18.00 ± 5.63 ^{abA}	18.22 ± 5.59 ^{abB}	19.59 ± 5.71 ^{bB}	20.15 ± 5.17 ^{bcB}	20.86 ± 4.41 ^{bcB}	23.24 ± 2.54 ^{cB}	18.92 ± 6.57 ^{B}

Values with equal letters, lowercase letters in the row and uppercase letters in the column, do not differ by Tukey test ($p > 0.05$). Data with normal distribution assessed by the Cramer–von Mises test at $p > 0.05$.

After three months (T3M) of evaluation, a progressive increase in Schirmer's test values was reported in the animals submitted to canthoplasty, with an improvement in ocular lubrication and a reduction in mucopurulent elasticity, due to a reduction in the palpebral fissure and less exposure of the eyeball.

In the group that underwent canthoplasty (treated group), there was correction of lagophthalmos and better eyelid closure, resulting in greater protection of the ocular surface and a reduction in the consequences of keratoconjunctivitis sicca [8,9]. In the control group, there was no significant improvement in ocular lubrication, even with the use of eye drops. Furthermore, in the surgically treated group, it was possible to extend the interval of use of tear drops, reducing the long-term efficiency of canthoplasty [9]. Clinical treatment alone was not effective in all cases, requiring a surgical approach to better lubricate the ocular surface [10].

Analysis of the TFBUT showed that the animals in the treated group showed a growing increase in TFBUT values over the evaluation period (between T0 and T18M) (Table 2).

Table 2. Tear film break-up time (TFBUT) in brachycephalic animals submitted to clinical treatment (control group) or early blepharoplasty (treated group) during the evaluation periods. Data are presented as mean ± standard deviation (S.D.).

Groups	Tear Film Break-Up Time (TFBUT) (s)									Mean ± S.D. Overtime
	T0	**T7**	**T30**	**T3M**	**T6M**	**T9M**	**T12M**	**T15M**	**T18M**	
Control group	6.73^{aA} ± 4.32	6.91^{aA} ± 4.44	7.00^{aA} ± 3.52	6.45^{aA} ± 3.69	5.90^{aA} ± 3.66	5.27^{aA} ± 3.08	3.43^{aA} ± 1.65	3.29^{aA} ± 1.64	3.29^{aA} ± 1.64	5.64^{A} ± 3.63
Treated group	6.38^{aA} ± 3.87	6.27^{aA} ± 3.74	6.17^{aA} ± 3.48	7.32^{abA} ± 2.98	7.80^{abB} ± 3.02	9.34^{bcB} ± 4.26	9.81^{cB} ± 3.42	10.14^{cB} ± 2.69	11.00^{cB} ± 3.41	7.78^{B} ± 3.79

Values with equal letters, lowercase letters in the row, and uppercase letters in the column do not differ by Tukey test ($p > 0.05$). Data with normal distribution assessed by the Cramer–von Mises test at $p > 0.05$.

The TFBUT revealed a significant improvement in tear film quality in the surgically treated group over time, with a progressive increase in values after the procedure. However, the mean values were below what is considered normal for brachycephalic breeds (>17 s) [11]. In the control group, a reduction in mean TFBUT values was observed, reducing the worsening of tear quality in dogs submitted to clinical treatment alone (Table 2).

In the present study, we observed that the eyes of the animals belonging to the treated group showed a reduction in the classification scores of the TCT (Figure 2, Table 3).

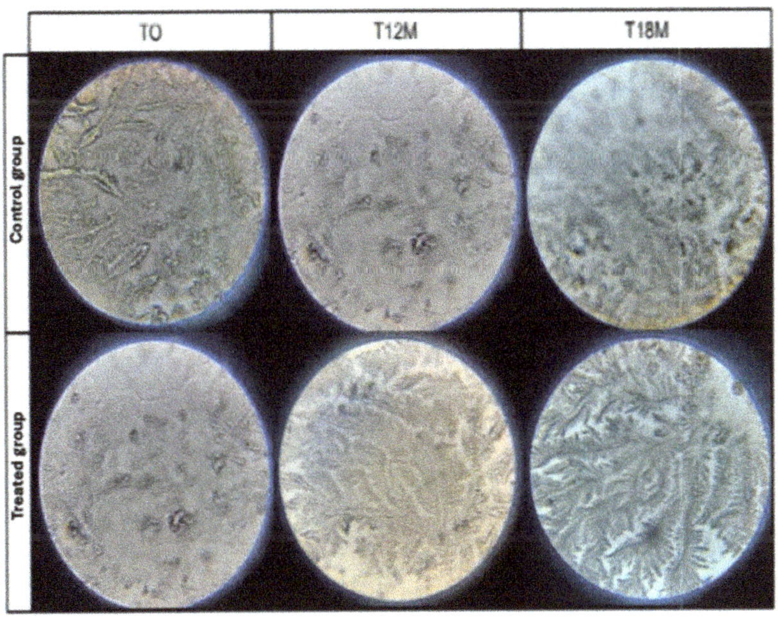

Figure 2. Representative images of the tear crystallization test (TCT) performed on brachycephalic canine patients before (T0) and after clinical (control group) or surgical treatment (treated group) at different times (T12M and T18M)—personal archive, 2021.

Table 3. Tear crystallization test (TCT) in brachycephalic animals submitted to clinical treatment (control group) or early blepharoplasty (treated group) during the evaluation periods. Data are presented as mean ± standard deviation (S.D.).

Groups	Tear Crystallization Test (TCT)									Mean ± S.D. Overtime
	T0	T7	T30	T3M	T6M	T9M	T12M	T15M	T18M	
Control group	2.82 ^{aA} ± 0.73	2.91 ^{aA} ± 0.68	2.91 ^{aA} ± 0.52	3.00 ^{aA} ± 0.62	3.00 ^{aA} ± 0.62	3.09 ^{aA} ± 0.52	3.29 ^{aA} ± 0.47	3.29 ^{aA} ± 0.47	3.29 ^{aA} ± 0.47	3.03 ^{A} ± 0.61
Treated group	2.90 ^{aA} ± 0.88	2.95 ^{aA} ± 0.84	2.94 ^{aA} ± 0.85	2.78 ^{aA} ± 0.69	2.78 ^{aA} ± 0.65	2.72 ^{aA} ± 0.75	2.56 ^{bB} ± 0.75	2.47 ^{bB} ± 0.75	2.29 ^{bB} ± 0.77	2.77 ^{B} ± 0.79

Values with equal letters, lowercase letters in the row and uppercase letters in the column do not differ, respectively, by the Kruskal–Wallis and Wilcoxon–Tukey tests ($p > 0.05$). Data with normal distribution assessed by the Cramer–von Mises test at $p > 0.05$.

In the control group, there was an increase in the TCT scores over the period evaluated, with a worsening in tear film quality, and a predominance of patterns III and IV, characteristic of dry keratoconjunctivitis [7,12]. In addition, the reduction in Schirmer's test values coincided with the worsening of tear quality, proving quantitative and qualitative dry keratoconjunctivitis.

In the control group, even with clinical treatment, the occurrence of corneal ulcers is frequent (Table 4).

Table 4. Frequency of occurrence and percentage of corneal ulcers in brachycephalic animals submitted to clinical treatment (control group) or early blepharoplasty (treated group) during the evaluation periods.

Groups	Presence or Absence of Corneal Ulcers	T7	T30	T3M	T6M	T9M	T12M	T15M	T18M
Control group	Presence	7 10.7%	0 0%	0 0%	3 4.8%	4 6.4%	3 4.8%	2 3.2%	4 6.4%
	Absent	15 23.0%	22 34.4%	22 34.4%	18 28.6%	17 27%	18 28.9%	19 30.2%	17 27%
Treated group	Present	14 21.5%	0 0%	0 0%	0 0%	0 0%	1 1.6%	1 1.6%	0 0%
	Absent	29 44.6%	42 65.65	42 65.6%	42 66.7%	42 66.7%	41 65.1%	41 65.1%	42 66.7%
Tests	Q-square (p value)	0.95	-	-	-	-	-	-	-
	Fisher test (p value)	-	1.00	1.00	0.03	0.01	0.09	0.21	0.01

Frequency of occurrence and percentage. Controls are 22 up to T3M and 21 from then onwards. Treated are 43 at T7 and 42 from then onwards. When observed frequency values are less than 5, use Fisher's test; for values greater than 5, use the Q-squared test. Frequencies are different when $p < 0.05$.

4. Discussion

Previous studies indicate that hereditary ophthalmic disorders in purebred dogs can appear from 6 weeks of age [13], and that puppies of brachycephalic breeds already show alterations due to ocular exposure [14]. Therefore, the inclusion of young animals was essential to evaluate preventive strategies for these lesions.

The ocular surface of brachycephalic dogs is highly susceptible to external injuries [15], and stimuli can induce greater tear production [16]. Processes caused by external agents, such as conjunctival hyperemia and edema, can rapidly develop into severe conditions [17].

Excessive ocular exposure in brachycephalic dogs is associated with ocular proliferation and the development of keratoconjunctivitis sicca [18]. Canthoplasty improves ocular lubrication by reducing corneal exposure and improving tear film distribution [19,20].

The persistence of conformational alterations in eyes treated only with topical therapy maintained poor lubrication in the central part of the cornea and could lead to subclinical inflammatory processes that impact tear production. The association of correction with clinical treatment has been shown to increase tear production in the medium term, providing better inflammatory control of the ocular surface and protective effects when performed early [21].

The TFBUT is an important diagnostic tool in the assessment of keratoconjunctivitis sicca and tear film stability in dogs [22,23]. Brachycephalic dogs are predisposed to tear film dysfunction due to exophthalmos and lagophthalmos, resulting in accelerated evaporation of the water layer [24,25]. Film rupture occurs when evaporation exceeds capillary flow, forming dark spots on the cornea [26]. Tear film instability has a direct impact on ocular health, influencing osmolarity and potentially leading to hyperosmolarity [27]. In humans, tear film osmolarity and the tear crystallization test (TCT) are used as markers of dry eye syndrome [28]. Although the TCT is used in veterinary ophthalmology [29,30], there are still no specific reports on brachycephalic breeds, and it was not measured in this study.

There was a statistically significant difference at T12M. It can therefore be inferred that there was an improvement in the quality of the tear film in the animals that underwent surgical correction in the medium and long term. According to Rolando [7] type I and II tears, which are related to a greater presence of crystals and less spacing between them, are tears of good quality, i.e., consistent with animals with a healthy ocular surface.

The TCT provides a qualitative assessment of the tear and may not be directly correlated with its quantity [31]. The interaction between electrolytes and macromolecules influences crystallization, and the absence of a halo suggests a reduction in the mucin layer and a possible decrease in goblet cells [32,33].

Ocular brachycephalic syndrome contributes to the early evaporation of the tear film and increased corneal exposure, favoring dehydration and ocular lesions [34,35]. Kerato conjunctivitis sicca can be aggravated by multiple factors, such as anatomical conformation, infections, palpebral disorders, and chronic irritations [36].

Conformational factors, such as a short snout, caruncle entropion, macrocephaly and lagophthalmos, increase the predisposition of brachycephalic dogs to corneal ulcers [3,11] observed that 81.18% of dogs with corneal ulcers were brachycephalic, reinforcing the influence of cranial morphology on their early occurrence.

Correcting anatomical abnormalities significantly reduces the risk of corneal ulcers in brachycephalic dogs [3,37]. They identified a predisposition to ulcerative keratitis, especially in young Shih Tzus (<3 years old). Recurrent ulcers occur when conformational changes persist.

Bedford & Jones [38] point out that macropalpebral fissures and shallow orbits expose the eyes of brachycephalic dogs, increasing the risk of corneal damage. Exophthalmos is also a predisposing factor for ulcerative keratitis, as it makes the ocular bulb more susceptible to trauma [37]. In addition, lagophthalmos, which is characteristic of brachycephalic ocular syndrome, compromises ocular lubrication and favors keratoconjunctivitis sicca. Eyelid and ciliary alterations, such as trichiasis and caruncle entropion, were observed in most of the animals evaluated. Caruncular trichiasis, when in constant contact with the cornea, causes chronic irritation and corneal hyperpigmentation [34,39].

Ulcerative keratitis can be associated with decreased tear production [40] or with low blink frequency and changes in tear film composition [41]. Lower corneal sensitivity in brachycephalics reduces the response to irritating stimuli and compromises tear distribution, favoring ocular lesions [42].

This study stands out among the existing literature for prospectively investigating the effects of early canthoplasty on the prevention of ophthalmic lesions in brachycephalic dogs for 18 months. Although previous studies [19] described post-surgical complications and the perception of tutors after canthoplasty, there is scant quantitative data regarding the clinical benefits obtained from this intervention [5]. Other authors [8,34] have highlighted the high prevalence of ocular lesions in brachycephalic breeds, but without exploring the impacts of an early surgical intervention. The present work contributes to filling this gap in the scientific knowledge by demonstrating that early canthoplasty performed before the development of irreversible lesions such as deep ulcers and pigmented keratitis significantly improves tear quality and reduces ocular complications. Consequently, the present results validate the efficacy of canthoplasty and support its preventive use. However, this study remains limited by its relatively low caseload, and larger studies are warranted to expand on this subject.

5. Conclusions

Overall, early canthoplasty in brachycephalic dogs promoted a reduction in the palpebral fissure, improved control of the quantity and quality of ocular lubrification, and prevented the mid-term effects of dry keratoconjunctivitis, supporting the study's hypothesis. Early canthoplasty was more beneficial than conventional clinical therapy in the mid-term. Adequate follow-up remains essential to perform tear secretion control and monitor ocular health.

Author Contributions: Conceptualization, A.L.A.-S.; methodology, A.R.C. and T.A.R.; formal analysis, A.R.C. and T.A.R.; investigation, A.R.C., T.A.R., D.M.C.S., J.R.d.S.J. and T.B.L.; resources, A.L.A.-S.; writing—original draft preparation, A.R.C. and A.I.F.-R.; writing—review and editing, R.M.G.d.C. and A.I.F.-R.; project administration, A.L.A.-S.; funding acquisition, R.M.G.d.C. and P.A.O. All authors have read and agreed to the published version of the manuscript.

Funding: This work was supported by CAPES, finance code 001. and by National Funds from the FCT–Portuguese Foundation for Science and Technology, for the projects UI/04033 and LA/P/0126/2020 (https://doi.org/10.54499/LA/P/0126/2020). Ana Lucia Abreu-Silva was a research productivity fellow of the National Scientific and Technological Development Council (Conselho Nacional de Desenvolvimento Científico e Tecnológico—CNPq) [grant number: 313348/2021-9].

Institutional Review Board Statement: This research was approved by the ethics committee (CEEA, protocol 35/2018) and carried out with the permission of the owners. Sixty-four eyes of brachycephalic dogs (6–24 months) treated at the Universidade Estadual do Maranhão (UEMA) Veterinary Hospital were analyzed.

Informed Consent Statement: Not applicable.

Data Availability Statement: Data available under request.

Conflicts of Interest: The authors declare no conflict of interest.

Abbreviations

The following abbreviations are used in this manuscript:

TCT	Tear crystallization test
TFBUT	Tear film break-up time

References

1. Nicholas, F.W.; Wade, C.M.; Williamson, P. Disorders in pedigree dogs: Assembling the evidence. *Vet. J.* **2010**, *183*, 8–9. [CrossRef] [PubMed]
2. Kobashigawa, K.K.; Lima, T.B.; Padua, I.R.M.; Sobrinho, A.A.F.d.B.; Marinho, F.d.A.; Ortêncio, K.P.; Laus, J.L. Ophthalmic parameters in adult Shih Tzu dogs. *Cienc. Rural* **2015**, *45*, 1280–1285. [CrossRef]
3. Packer, R.M.A.; Hendricks, A.; Tivers, M.S.; Burn, C.C.; Ambrósio, C.E. Impact of facial conformation on canine health: Brachycephalic obstructive airway syndrome. *PLoS ONE* **2015**, *10*, e0137496. [CrossRef]
4. Gelatt, K.N.; Gelatt, J.P. Surgery of the cornea and sclera. In *Veterinary Ophthalmic Surgery*, 1st ed.; Gelatt, K.N., Gelatt, J.P., Eds.; Elsevier Health Sciences: St. Louis, MO, USA, 2011.
5. Andrews, A.L.M.M.; Youngman, K.L.; Packer, R.M.A.; O'neill, D.G.; Kafarnik, C. A review of clinical outcomes, owner understanding and satisfaction following medial canthoplasty in brachycephalic dogs in a UK referral setting (2016–2021). *Animals* **2023**, *13*, 2032. [CrossRef] [PubMed]
6. Gelatt, K.N.; Gelatt, J.P. *Veterinary Ophthalmic Surgery*; Saunders: St. Louis, MO, USA, 1994.
7. Rolando, M. Tear mucus ferning test in normal and keratoconjunctivitis sicca eyes. *Chibret Int. J. Ophthalmol.* **1984**, *2*, 32–41.
8. Pe'er, O.; Ofri, R.; Sebbag, L. Schirmer tear test-1 with open or closed eyelids: An evaluation in brachycephalic and nonbrachycephalic dogs. *Vet. Ophthalmol.* **2025**, *28*, 21–27. [CrossRef]
9. Gelatt, K.N. *Essentials of Veterinary Ophthalmology*, 2nd ed.; Blackwell Publishing Professional: Ames, IA, USA, 2008; p. 81.
10. Barachetti, L.; Rampazzo, A.; Mortellaro, C.M.; Scevola, S.; Gilger, B.C. Use of episcleral cyclosporine implants in dogs with keratoconjunctivitis sicca: Pilot study. *Vet. Ophthalmol.* **2015**, *18*, 234–241. [CrossRef]
11. Santos, T.G.S. Incidência de Ceratite Ulcerativa em Cães–Estudo Comparativo em Braquicefálicos e Não Braquicefálicos. Bachelor's Thesis, CEPLAC–Faculdade de Medicina Veterinária, Itabuna, Brazil, 2020.
12. Masmali, A.M.; Al-Qhtani, S.; Al-Gasham, T.M.; El-Hiti, G.A.; Purslow, C.; Murphy, P.J. Application of a new grading scale for tear ferning in non-dry eye and dry eye subjects. *Cont. Lens Anterior Eye* **2015**, *38*, 39–43. [CrossRef]
13. Belknap, E.; Genetics Committee of the American College of Veterinary Ophthalmologists. Ocular disorders, proven or presumed to be inherited in purebred dogs. In *The Blue Book: Ocular Disorders Presumed to Be Inherited in Purebred Dogs*, 13th ed.; ACVO, Ed.; American College of Veterinary Ophthalmologists: Meridian, ID, USA, 2015.
14. Costa, J.V. Síndrome Braquicefálica Ocular. Master's Thesis, Universidade de Lisboa, Lisbon, Portugal, 2019.
15. Gipson, I.K. The ocular surface: The challenge to enable and protect vision. *Investig. Ophthalmol. Vis. Sci.* **2007**, *48*, 4391–4398. [CrossRef]
16. Christmas, R. Common ocular problems of Shih Tzu dogs. *Can. Vet. J.* **1992**, *33*, 390–393.
17. Galley, A.P.; Beltran, E.; Tetas Pont, R. Neurogenic keratoconjunctivitis sicca in 34 dogs: A case series. *Vet. Ophthalmol.* **2022**, *25*, 140–152. [CrossRef]
18. Cooper, S. Keratoconjunctivitis sicca in the dog. *Companion Anim.* **2012**, *17*, 37–42. [CrossRef]
19. Allgoewer, I. Simplified medial canthoplasty: Technique and early postoperative complications in 601 dogs (1180 eyes). *Vet. Ophthalmol.* **2025**, *28*, 497–505. [CrossRef]
20. Appelboam, H. Pug appeal: Brachycephalic ocular health. *Companion Anim.* **2016**, *1*, 36. [CrossRef]
21. Milner, M.S.; Beckman, K.A.; Luchs, J.I.; Allen, Q.B.; Awdeh, R.M.; Berdahl, J.; John, B.; Boland, T.S.; Buznego, C.; Gira, J.P.; et al. Dysfunctional tear syndrome: Dry eye disease and associated tear film disorders. *Curr. Opin. Ophthalmol.* **2017**, *28*, 3–47. [CrossRef]
22. Johnson, M.E.; Murphy, P.J.; Boulton, M. Effectiveness of sodium hyaluronate eyedrops in the treatment of dry eye. *Graefe's Arch Clin Exp Ophthalmol.* **2006**, *244*, 109–112. [CrossRef] [PubMed]
23. Cullen, C.L.; Ihle, S.L.; Webb, A.A. Keratoconjunctival effects of diabetes mellitus in dogs. *Vet. Ophthalmol.* **2005**, *8*, 215–224. [CrossRef]
24. Gelatt, K.N.; Gilger, B.C.; Kern, T.J. *Veterinary Ophthalmology*, 5th ed.; Wiley-Blackwell: Hoboken, NJ, USA, 2013.
25. Pigatto, J.A.T.; Pereira, F.Q.; Almeida, A.C.V.R.; Redaeli, R.; Faganello, C.S.; Franzen, A.A. Keratoconjunctivitis in dogs and cats. *Acta Sci. Vet.* **2007**, *35*, 250–251.
26. Peng, C.-C.; Cerretani, C.; Braun, R.J.; Radke, C. Evaporation-driven instability of the precorneal tear film. *Adv. Colloid Interface Sci.* **2014**, *206*, 250–264. [CrossRef]
27. Murube, J. Tear osmolarity. *Ocul. Surf.* **2008**, *4*, 62–73. [CrossRef] [PubMed]
28. Baudouin, C.; Aragona, P.; Messmer, E.M.; Tomlinson, A.; Calonge, M.; Boboridis, K.G.; Akova, Y.A.; Geerling, G.; Labetoulle, M.; Rolando, M. Role of hyperosmolarity in dry eye disease. *Ocul. Surf.* **2013**, *11*, 246–258. [CrossRef] [PubMed]
29. Mota, L.P.A. Evaluation of the Ocular Surface of Dogs with Protrusion of the Third Eyelid Gland. Ph.D. Thesis, Universidade de Brasília, Brasília, Brazil, 2014.
30. Silva, L.R.; Gouveia, A.F.; de Fátima, C.J.; Oliveira, L.B.; Reis, J.L., Jr.; Ferreira, R.F.; Pimentel, C.M.; Galera, P.D. Tear ferning test in horses and its correlation with ocular surface evaluation. *Vet. Ophthalmol.* **2016**, *19*, 117–123. [CrossRef]

31. Avancini, J.B. Tear Crystallization Test in Healthy Shih Tzu Dogs. Bachelor's Thesis, Faculdade de Medicina Veterinária da FAV, Brasília, Brazil, 2016.
32. McGinnigle, S.; Naroo, S.A.; Eperjesi, F. Evaluation of dry eye. *Surv. Ophthalmol.* **2012**, *57*, 293–316. [CrossRef]
33. Julio, G.; Lluch, S.; Pujol, P.; Merindano, M.D. Effects of tear hyperosmolarity on conjunctival cells. *Ophthalmic Physiol. Opt.* **2012**, *32*, 317–323. [CrossRef]
34. Sebbag, L.; Sanchez, R.F. The pandemic of ocular surface disease in brachycephalic dogs: The brachycephalic ocular syndrome. *Vet. Ophthalmol.* **2023**, *26*, 31–46. [CrossRef]
35. Lima, A.M.V. Tear Production and Goblet Cell Density in Shih-Tzu. Ph.D. Thesis, Universidade Federal de Goiás, Goiânia, Brazil, 2008.
36. Iwashita, H.; Sebbag, L.; Leonard, B.C.; Saito, A. A review of diagnostic tests for qualitative and quantitative tear film deficiency in dogs. *Vet. Ophthalmol.* **2023**, *26*, 5–15. [CrossRef]
37. Kim, J.Y.; Won, H.J.; Jeong, S.W. A retrospective study of ulcerative keratitis in 32 dogs. *Int. J. Appl. Res. Vet. Med.* **2009**, *7*, 27–31.
38. Bedford, P.G.C.; Jones, R.G. Abnormal appearance. In *Small Animal Ophthalmology: A Problem-Oriented Approach*; Peiffer, R.L., Jr., Petersen-Jones, S.M., Eds.; W.B. Saunders: Philadelphia, PA, USA, 2001; pp. 94–95.
39. Voitena, J.N.; Brito, F.L.C.; Marinho, T.O.C.; Montiani-Ferreira, F.; Cremonini, D.N.; Chiurciu, J.L.V.; Jesus, N.S.; Leonard, B.C.; da Cunha, O. Application of OSA-VET® and qualiquantitative tear tests in brachycephalic dogs with and without keratoconjunctivitis sicca. *Vet. Res. Commun.* **2025**, *49*, 40. [CrossRef] [PubMed]
40. Imcharoon, K.; Pinyosnit, N.; Srilert, P.; Ngampongsai, T.; Hunprasit, V.; Tuntivanich, N.; Jaturakan, O. Comparative study of healing time of canine non-infectious deep ulcerative keratitis between medical therapy alone and combined treatment with medical therapy and a nictitating membrane flap: A retrospective study. *Open Vet. J.* **2022**, *12*, 815–821. [CrossRef] [PubMed]
41. Carrington, S.D.; Bedford, P.G.; Guillon, J.P.; Woodward, E.G. Biomicroscopy of the tear film of the Pekingese dog. *Vet. Rec.* **1989**, *124*, 323–328. [CrossRef]
42. Startup, F.G. Corneal ulceration in the dog. *J. Small Anim. Pract.* **1984**, *25*, 737–752. [CrossRef]

veterinary sciences

MDPI

Article

Evaluation of the Dual Antiviral and Immunomodulatory Effects of *Phallus indusiatus* in a Feline Infectious Peritonitis Model Using PBMCs

Chularat Hlaoperm [1,2], Wassamon Moyadee [1,2], Emwalee Wongsaengnoi [3], Wiwat Klankaew [4], Amonpun Rattanasrisomporn [5], Atchara Paemanee [6,7], Kiattawee Choowongkomon [4], Oumaporn Rungsuriyawiboon [3] and Jatuporn Rattanasrisomporn [1,2,*]

1 Graduate Program in Animal Health and Biomedical Sciences, Faculty of Veterinary Medicine, Kasetsart University, Bangkok 10900, Thailand; chularat.h@ku.th (C.H.); wassamon.moy@ku.th (W.M.)
2 Department of Companion Animal Clinical Sciences, Faculty of Veterinary Medicine, Kasetsart University, Bangkok 10900, Thailand
3 Department of Veterinary Technology, Faculty of Veterinary Technology, Kasetsart University, Bangkok 10900, Thailand; emwalee.wo@ku.th (E.W.); cvtopr@ku.ac.th (O.R.)
4 Department of Biochemistry, Faculty of Science, Kasetsart University, Bangkok 10900, Thailand; wiwat.klankaew@gmail.com (W.K.); fsciktc@ku.ac.th (K.C.)
5 Interdisciplinary of Genetic Engineering and Bioinformatics, Graduate School, Kasetsart University, Bangkok 10900, Thailand; fgraapr@ku.ac.th
6 National Omics Center, National Center for Genetic Engineering and Biotechnology (BIOTEC), National Science and Technology Development Agency (NSTDA), Khlong Luang, Pathum Thani 12120, Thailand; atchara.pae@biotec.or.th
7 Food Biotechnology Research Team, Functional Ingredients and Food Innovation Research Group, National Center for Genetic Engineering and Biotechnology (BIOTEC), National Science and Technology Development Agency (NSTDA), Khlong Luang, Pathum Thani 12120, Thailand
* Correspondence: fvetjpn@ku.ac.th

Simple Summary: This study explores the potential of *Phallus indusiatus* (*P. indusiatus*), a medicinal mushroom, as a dual-action treatment for feline infectious peritonitis (FIP), a fatal disease caused by feline coronavirus. *P. indusiatus* demonstrated the most pronounced inhibitory effect among the 17 mushroom extracts tested (69.2% inhibition of FIPV M^{pro}), showing a level of activity approaching that of standard antivirals such as lopinavir and ritonavir. The extract also significantly reduced nitric oxide production (a marker of inflammation) in lipopolysaccharide (LPS)-stimulated peripheral blood mononuclear cells (PBMCs) from healthy cats, indicating anti-inflammatory effects. However, immune cells from cats already suffering from FIP, which were in a chronically inflamed state, did not respond to the extract or to dexamethasone. This highlights the complexity of FIP-related inflammation. This study also successfully cultured PBMCs from FIP fluid for the first time, providing a novel in vitro model for future research. Overall, the results suggest that *P. indusiatus* has promising antiviral and anti-inflammatory properties, though further research is needed to isolate active compounds and test in live infection models.

Abstract: Feline infectious peritonitis (FIP) is a progressive and often fatal disease caused by a virulent biotype of feline coronavirus (FCoV). Although antiviral treatments are now available, relapse and resistance remain ongoing concerns. This study investigates the therapeutic potential of *P. indusiatus*, a medicinal mushroom, for its antiviral and anti-inflammatory activities against FIP. The main protease (FIPV M^{pro}) of feline infectious peritonitis virus (FIPV) was recombinantly expressed and purified to facilitate enzyme inhibition screening. *P. indusiatus* exhibited the strongest FIPV M^{pro} inhibitory activity among the 17 mushroom extracts tested (69.2%), showing a notable level of inhibition relative to standard antiviral agents such as lopinavir and ritonavir. To assess its anti-inflammatory potential, PBMCs derived from healthy cats and FIP-associated effusions (FIP fluid) were cultured and stimulated with LPS to induce inflammation. In healthy PBMCs, *P. indusiatus* significantly reduced nitrite levels, with effects similar to dexamethasone. However, PBMCs from FIP fluid, already in an activated state, showed no additional response. Notably, this study is the first to successfully isolate and culture PBMCs from FIP fluid, providing a new platform for future immunological research. These findings suggest that *P. indusiatus* possesses both antiviral and anti-inflammatory properties, positioning it as a potential dual-action therapeutic candidate for FIP. Further investigation into cytokine signaling pathways is warranted to clarify its mechanisms of action and advance future therapeutic development.

Keywords: *Phallus indusiatus*; PBMCs; feline coronavirus (FCoV); feline infectious peritonitis (FIP); FIPV protease; FIPV M^{pro}; antiviral; anti-inflammatory

1. Introduction

Feline coronavirus (FCoV) is an enveloped, positive-sense single-stranded RNA virus in the Coronaviridae family commonly found in domestic cats [1]. It exists in two biotypes: feline enteric coronavirus (FECV) [2], which typically causes mild gastrointestinal signs and is often self-limiting in immunocompetent cats, and feline infectious peritonitis virus (FIPV), the pathogenic variant responsible for feline infectious peritonitis (FIP) [3]. First reported in 1966, FIP is fatal, immune-mediated disease that primarily affects kittens, adolescents, and young adult cats, and remains a significant cause of mortality in this population [4]. Approximately 12% of FCoV-infected cats progress to FIP, with environmental stress and immune dysfunction as a key contributing factors [5]. Although antiviral drugs have recently become available, delayed diagnosis still results in high mortality, emphasizing the need for timely and accurate detection [6]. Pathogenesis involves efficient replication of FCoV in activated monocytes and macrophages, leading to systemic inflammation, pyogranulomatous lesions, and T-cell depletion [6,7]. Altered cytokine profiles, including elevated interleukin (IL)-1β, IL-6, IL-10, IL-12p40, and tumor necrosis factor-alpha (TNF-α), have been identified in affected cats, indicating that immune dysregulation plays a key role in the advancement of FIP [8].

Edible mushrooms have been reported to possess multiple biomedical properties, including antiviral, anti-inflammatory activities, and immunomodulatory effects through cytokine stimulation [7,8]. These effects are largely attributed to bioactive compounds such as polysaccharides, terpenoids, and alkaloids. These compounds enhance phagocytic activity, increase IL-1β and TNF-α expression, and activate key signaling pathways such as ERK1/2, JNK, and p38 MAPKs. They also promote NF-κB p65 phosphorylation and nuclear translocation and can bind to macrophages via Toll-like receptor 4 (TLR4) [8]. Certain mushrooms may also help alleviate clinical signs associated with cytokine storms. While some compounds have shown antiviral activity, scientific evaluation in feline models remains limited.

In the preliminary inflammation model of this study, both lipopolysaccharide (LPS) and FIPV (or viruses in general) are potent activators of the NF-κB and AP-1 signaling pathways [9–11]. These pathways function as central molecular "switches" that regulate the transcription of various inflammatory mediators, including NO, inducible nitric oxide synthase (iNOS), cyclooxygenase-2 (COX-2), and prostaglandin E2 (PGE2), and the pro-inflammatory cytokines (e.g., TNF-α, IL-6, and IL-1β) [12,13]. Therefore, using LPS as an inflammatory stimulant in PBMCs to investigate the anti-inflammatory effects of mushroom extracts is well justified as a preliminary inflammation model, as it activates shared pathways involved in cytokine and iNOS expression mechanisms that are also triggered by FIPV (Figure 1).

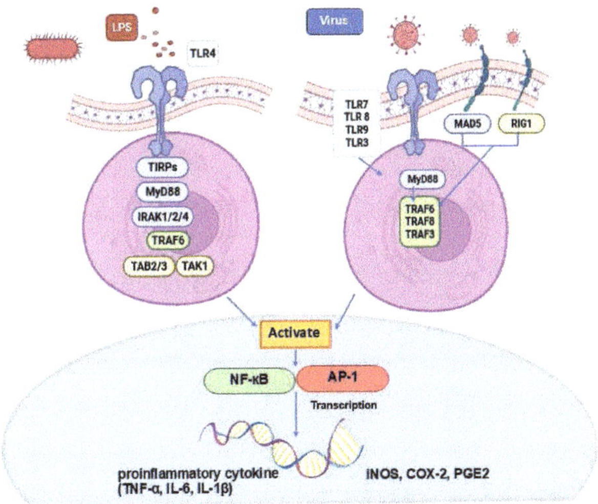

Figure 1. This figure illustrates the activation of TLR–NF-κB and AP-1 signaling pathways in response to LPS or viral stimulation, resulting in the upregulation of inducible iNOS, COX-2, PGE2, and pro-inflammatory cytokines. This schematic was adapted from Luo R. et al., 2025, Yu, L. et al., 2024, and Hennessy & McKernan., 2021 [9–11].

Coronaviruses are RNA viruses that hijack the host's translational machinery to generate viral proteins. The resulting polyproteins must be cleaved to release individual functional proteins necessary for viral replication and transcription. The virus encodes proteases, including the main protease (M^{pro}), also known as 3CLpro, which has conserved sequences that facilitate the assembly of the viral replicase complex [14]. The M^{pro} protein of FCoV is essential for the replication cycle, as observed in other members of the Coronaviridae family [15]. Therefore, FIPV M^{pro} protein is considered a prominent drug target for antiviral therapy against FIP (e.g., GC376) [14,16].

The immune system plays a crucial role in controlling viral infections, including the suppression of FCoV replication and prevention of progression to FIP. In this study, 17 mushroom extracts were screened as potential candidates for alternative immunomodulatory strategies in cats. Screening was conducted to assess antiviral activity by them against FIPV M^{pro} to identify active compounds. The extract with the strongest inhibitory effect was selected for further assessment of its anti-inflammatory potential. PBMCs were isolated from both healthy cats and cats with FIP-associated effusions, including pleural, peritoneal, or a combination of both, and were cultured under optimized conditions. These PBMCs were then stimulated with LPS and treated with the selected extract to assess nitrite-scavenging activity. The resulting data will serve as a foundation for future investigations into cytokine modulation, particularly in the context of FIP-related inflammation and antiviral response.

2. Materials and Methods

2.1. Ethics Statement

This study was approved by the Institutional Animal Care and Use Committee of Kasetsart University, Bangkok, Thailand, under protocol number ACKU67-VET-124. Additionally, all sample collections were performed with the informed consent of the cat owners.

2.2. Protein Expression and Purification

The expression and purification of FIPV's main protease in this study, called "FIPV M^{pro}", have been described previously [17]. In this study, the FIPV M^{pro} was derived from the FIPV WSU-79-1146 strain. The coding sequence for M^{pro} included an N-terminal 10xHis tag and a TEV cleavage site, resulting in a recombinant protein with an approximate molecular weight of 37.2 kDa. The gene was synthesized into the pGEX-4T-1 vector (GenScript, USA) and transformed into *Escherichia coli* strain BL21(DE3) for protein expression. Cultures were grown in LB medium incubated at 37 °C for 3 to 4 h, the absorbance was measured at OD 600 to 0.8, and then the sample was induced by 0.4 mM isopropyl-β-d-thiogalactopyranoside at 37 °C for 18–20 h. The FIPV M^{pro} was purified by Ni^{2+} affinity column (Cytiva AB, Sweden). Western blotting was performed to confirm the protein expression, using an anti-His tag primary antibody (Thermo Fisher Scientific, USA) and an HRP-conjugated Goat Anti-Mouse IgG secondary antibody (Thermo Fisher Scientific, USA) for signal detection.

2.3. Inhibition of FIPV M^{pro} Protease

The FIPV M^{pro} modified with Dabcyl-KTSAVLQSGFRKM-E (Edans) (Genscript USA, Piscataway, NJ, USA) was prepared in a buffer containing 20 mM Tris (pH 7.5), 100 mM NaCl, 2 mM DTT, and 0.05 mM EDTA. The enzyme activity was subsequently examined in a 384-well plate containing 1 μM FIPV M^{pro} in buffer solution. Subsequently, enzyme activity inhibitors, including lopinavir and itonavir as positive controls, were added for comparison with 10 μg/mL mushroom extracts. The final inhibitor concentration in the reaction was set to 10 μg/mL. The reaction was initiated by adding the substrate at a final concentration of 40 μM. The increase in fluorescence intensity of the substrate at 340 nm was measured to determine the relationship between Relative Fluorescence Units (RFUs) per second, using a microplate reader (Infinite 200, Tecan Group Ltd., Männedorf, Switzerland) and GraphPad version 8.0 software, as described in Equation (1).

$$\% \text{ Relative inhibition } = \text{V}_0 \text{ enz} - \left(\frac{\text{V}_0 \text{ Inhibitor} \times 100}{\text{V}_0 \text{ enz}} \right) \quad (1)$$

V_0 represents the initial reaction velocity of the enzyme. It indicates how fast the enzyme converts substrate to product at the start of the reaction. V_0 enz is the initial reaction velocity of the enzyme without an inhibitor. V_0 Inhibitor is the initial reaction velocity of the enzyme in the presence of an inhibitor.

Given the constraints associated with cell culture systems and the limited availability of animal-derived samples, particularly those obtained from experimental models, this phase of the study was designed to reduce animal usage by utilizing an in vitro screening approach. This model was employed to identify mushroom extracts exhibiting the highest inhibitory activity against the FIPV M^{pro}. The extract demonstrating the most potent inhibitory effect will be selected for further investigation of its anti-inflammatory properties.

2.4. Animals and Experimental Design

This research was approved for sample collection from the Kasetsart University Veterinary Teaching Hospital in Bangkok, Thailand. PBMCs were isolated from healthy donor cats ($n = 20$). In parallel, pleural and/or ascitic fluid samples were collected from cats clinically diagnosed with effusive FIP, based on a combination of consistent clinical, laboratory, and imaging findings. These included serum biochemical and hematological abnormalities, clinical signs such as anorexia, depression, intermittent fever, abdominal distension, or dyspnea due to pleural effusion. Radiographic or ultrasonographic imaging confirmed the presence of fluid accumulation in body cavities and involvement of internal organs. Fluid analysis and cytology revealed pyogranulomatous inflammation, with a proteinaceous, granular background, and a cell population consisting of macrophages, non-degenerate neutrophils, and few lymphocytes. The effusions were classified as modified transudates or non-septic exudates. The diagnosis was further supported by additional findings, including a serum albumin-to-globulin (A:G) ratio of less than 0.8, a positive Rivalta's test, and/or a positive PCR result for FCoV, in accordance with previously published criteria [18,19]. The pleural and/or ascitic fluid samples were subsequently used for mononuclear cell isolation. These samples ($n = 20$) were subsequently used for mononuclear cell isolation. For simplicity, mononuclear cells isolated from effusions are referred to as "PBMCs from FIP fluid" throughout this study. The sample size was derived from data in other similar studies. All analyses were performed using GraphPad Prism version 9 (GraphPad Software, LLC, Boston, MA, USA) [20].

2.5. The Preparation of the Mushroom Extraction Procedure

A total of 17 types of medical mushrooms were selected, including *Pleurotus ostreatus, Flammulina filiformis, Agrocybe cylindracea, Lentinula edodes, Tuber Uncinatum, Lignosus rhinocerotis, Volvariella volvacea, Ganoderma oregonense, Ganoderma formosanum, Ganoderma lucidum, Inonotus obliquus, Hericium erinaceus, Psilocybe cubensis, Phellinus igniarius, Dasoclema simensis, Pleurotus ostreatus*, and *Phallus indusiatus*. A dry sample of 100 g was air-dried and grinded by mixer grinder, immersed in 95% ethyl alcohol (ethanol) 200 mL at a temperature of 37 °C, and shaken for 12 h. Afterward, the liquid part was separated and the ethanol was evaporated using a rotary evaporator under vacuum. The crude extract was stored at a temperature of −20 °C until analysis. The extracts from the 17 types of mushrooms will be further tested for their FIPV M^{pro} inhibitory activity. In this study, crude extracts were prepared using the same stock and extraction protocol described by Theeraraksakul, K. et al., 2023 [21].

2.6. Preparation of PBMCs

Whole blood from healthy donor cats and effusion samples (pleural, ascitic, or both) from FIP cats were processed to isolate PBMCs by separating the buffy coats using centrifugation with Lymphoprep™ (Corning®, USA) at 600 g for 30 min to isolate PBMCs. Subsequently, complete RPMI-1640 medium (Corning®, USA) containing 10% fetal bovine serum (FBS; Gibco®, USA) and 1% penicillin–streptomycin (Gibco®, USA) was added, and the cell concentration was adjusted to 1×10^6 cells/mL. The cells were counted, and their viability was assessed by mixing the separated cells with a 0.4% Trypan Blue solution (Gibco®, USA) in a 1:1 ratio. A hemocytometer was used to count the cells under a microscope.

In the case of isolating PBMCs from effusion samples, referred to in this study as "PBMCs from FIP fluid" the sample characteristics of the FIP fluid were carefully considered. If the sample exhibited high viscosity, it was diluted with 1× phosphate-buffered saline (PBS; Cytiva HyClone™, USA) to facilitate efficient cell pelleting. The diluted sample was then centrifuged at 600× g for 10 min. The supernatant was discarded, and the cell pellet was retained. The pellet was subsequently washed three times with RPMI medium at a 1:1 ratio, each wash involving centrifugation at 600× g for 10 min. Once the cells were isolated, cell counting was performed, and the cells were cultured under the same conditions as previously described.

2.7. In Vitro Optimize Toxicity of P. indusiatus, LPS, and DMSO

The toxicity of *P. indusiatus* extract at concentrations of 10,000, 5000, 2500, 625, 156, 78, 39, and 19 µg/mL was tested on PBMCs isolated from the whole blood of donor cats. PBMCs (1×10^6 cells/mL) were seeded into a 96-well plate and pre-incubated for 1–2 h in a humidified atmosphere with 5% CO_2 at 37 °C. Subsequently, the cells were treated with various concentrations of *P. indusiatus* extract dissolved in 0.5% DMSO. The PBMCs were incubated for 24 h. Following treatment, cell viability was assessed using the Cell Counting Kit-8 (CCK-8) colorimetric assay (Dojindo, Kumamoto, Japan). Briefly, 10 µL of CCK-8 solution was added to each well and incubated for 2–4 h. The absorbance of each well was then measured at 450 nm using a microplate reader (EnSight® Multimode Plate Reader, PerkinElmer, Waltham, MA, USA). Cell viability was determined by calculating the percentage of absorbance relative to the control group (non-treated PBMCs).

The testing involved determining optimal concentrations and exposure times in generating PBMCs after stimulation with LPS (Sigma-Aldrich®, Israel) at various concentrations (0.001, 0.01, 0.1, and 1 µg/mL). Additionally, testing was conducted to determine suitable concentrations for cell viability using dimethyl sulfoxide (DMSO, Uvasol, Japan) at concentrations of 0.5%, 1%, 1.5%, and 2%, compared to a control group without DMSO. Cell viability was assessed using the CCK-8 colorimetric assay (Dojindo, Kumamoto, Japan) under the same conditions as described previously.

2.8. Anti-Inflammatory Assay Using Nitrite Scavenging Activity

To determine the effect of *P. indusiatus* extract on nitrite production, a primary breakdown product of nitric oxide, cells were suspended in a 96-well plate containing complete RPMI-1640 medium and pre-incubated for 1–2 h in a humidified atmosphere with 5% CO_2 at 37 °C. Subsequently, the cells were treated with *P. indusiatus* at a concentration of 19 µg/mL and dexamethasone at 5 µg/mL as a positive drug control in the presence of LPS (1 µg/mL) and incubated for 24 h. Nitrite production was determined by calculating the percentage of absorbance relative to the control group (non-treated PBMCs) using the Griess Reagent System (Promega, USA). A parallel test was conducted on PBMCs isolated from FIP fluid under the same conditions, but without LPS stimulation. This study investigates the potential effects of *P. indusiatus* extract on inflammation in PBMCs isolated from whole blood and fluid samples of cats diagnosed with FIP. The primary aim was to evaluate whether the extract can modulate inflammation by measuring nitrite concentrations, which is a marker of inflammatory response.

2.9. Statistical Analysis

Data are presented as mean ± Standard Error of the Mean (SEM), with each experiment replicated at least twice to ensure reliability. PBMC viability from whole blood and FIP fluid was compared using two-way ANOVA, with comparisons made between cell means within each row ($p < 0.05$, $n = 5$). The effect of varying LPS concentrations (0.001, 0.01, 0.1, and 1 µg/mL) on nitrite levels was assessed after 24 h using one-way ANOVA, compared to a no-LPS control ($p < 0.05$). The impact of 19 µg/mL *P. indusiatus* and 5 µg/mL dexamethasone on nitrite production in PBMCs from donor cat whole blood was evaluated after 24 h, with data shown as mean ± SEM ($n = 8$), comparing column means. Inflammation modulation in PBMCs from FIP fluid treated with *P. indusiatus* extract ($n = 5$) was analyzed using one-way ANOVA ($p < 0.05$). Differences in treatment responses between PBMCs from whole blood and FIP fluid were assessed by two-way ANOVA ($p < 0.05$, $n = 4$).

3. Results

3.1. Expression and Purification of FIPV M^{pro}

Recombinant His-tagged FIPV 3CLpro, expressed in *E. coli* BL21 (DE3), was purified using a Ni^{2+} affinity column. The expression of the FIPV M^{pro} or FIPV-3C-like Protease, ($3CL^{pro}$), was induced with 0.4 mM isopropyl β-d-1-thiogalactopyranoside (IPTG), followed by incubation for 18–24 h. Quantitative analysis using SDS-PAGE confirmed the expression of the FIPV M^{pro} protein with a molecular weight of 37.20 kDa (Figure 2A).

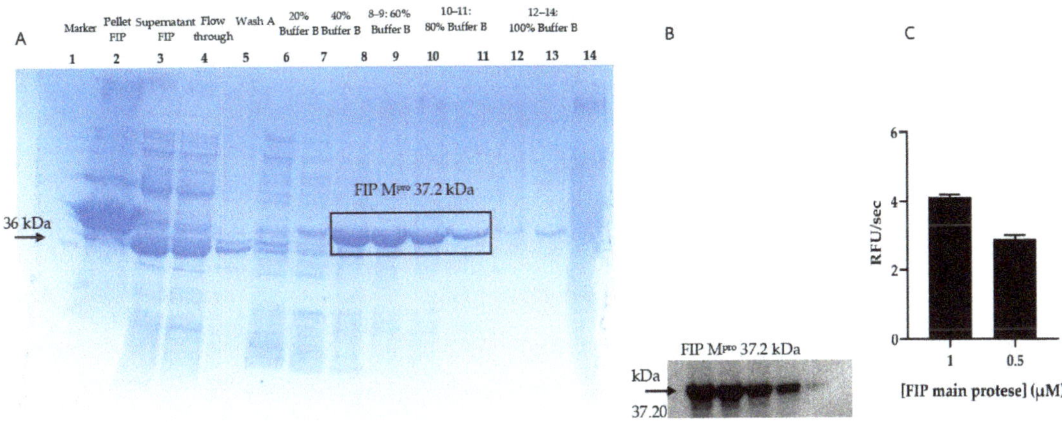

Figure 2. Protein purification was performed using a Ni^{2+} affinity column, with protein expression induced by 0.4 mM IPTG. The FIPV M^{pro} was detected at 37.2 kDa on an SDS-PAGE gel. lane1: Marker, lane2: Pellet FIP, lane3: Supernatant FIP, lane4: Flowthrough, lane5: Wash A, lane6: 20% Buffer B, lane7: 40% Buffer B, lane 8–9: 60% Buffer B, lane10–11: 80% Buffer B, lane12–14: 100% Buffer B (**A**). Western blot analysis confirmed the presence of the FIPV M^{pro} protein at 37.20 kDa (**B**). The reaction rates of the FIPV M^{pro} enzyme (Relative Fluorescence Units/sec) at enzyme concentrations of 1 and 0.5 μM (**C**).

Qualitative analysis was performed using the Western blot technique, which utilized the specificity of the primary antibody (Anti-His Tag (HRP)) for proteins labeled with a 6x His-Tag. A secondary antibody (Goat Anti-Mouse IgG (HRP)) was employed to bind with HRP substrates. This revealed a band corresponding to the FIPV M^{pro} protein at a size of 37.20 kDa (Figure 2B). The results of determining the optimal concentration of FIPV M^{pro} enzymatic activity, using Dabcyl-KTSAVLQSGFRKM-E (Edans) as the substrate at a final concentration of 40 μM, showed that the reaction rate (or fluorescence intensity) of the substrate increased as the enzyme concentration was raised to 0.5 and 1 μM, separately (Figure 2C).

3.2. Inhibition of FIPV M^{pro} Activity by Antiviral Drugs and Crude Mushroom Extract

The effect of various inhibitors on FIPV M^{pro} activity at a concentration of 1 μM was evaluated using protease inhibitors or antiviral drugs, with lopinavir and ritonavir (National Cancer Institute (NCI, USA)) serving as positive controls for comparison with mushroom extracts. Lopinavir and ritonavir were used at final concentrations of 10 and 100 μg/mL to assess enzyme inhibition by monitoring the reaction rate. This was performed by measuring the fluorescence intensity over 30 min. The initial reaction rate (V_0) was then used to calculate the percentage of enzyme activity inhibition. At a concentration of 10 μg/mL, lopinavir inhibited enzyme activity by 84.13%, while ritonavir inhibited it by 71.88% (Figure 3).

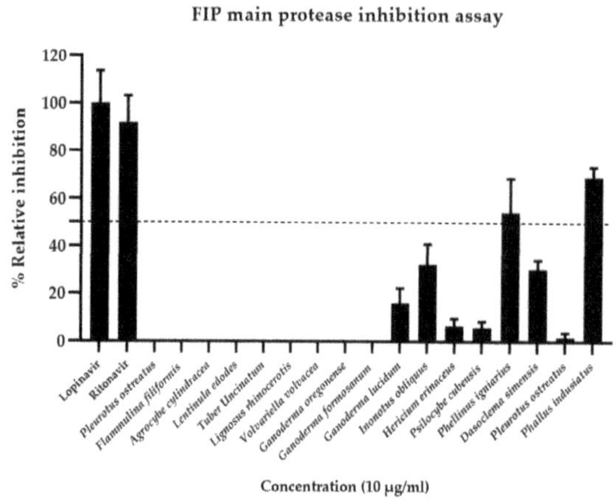

Figure 3. The relationship between the percentage inhibition of FIPV M^{pro} (1 μM) and crude mushroom extracts at a concentration of 10 μg/mL was evaluated. The inhibition of FIPV M^{pro} enzyme activity was evaluated at a concentration of 1 μM. At a concentration of 10 μg/mL, lopinavir inhibited enzyme activity by 84.13%, while ritonavir inhibited it by 71.88%. *P. indusiatus* exhibited the highest inhibition of FIPV M^{pro} activity at 69.2%. The dotted line represents 50% relative inhibition.

Relative inhibition of the FIPV M^{pro} assay was evaluated using crude mushroom extracts. The experimental results showed that *P. indusiatus* exhibited the highest inhibition of FIPV M^{pro} activity at 69.2%. Therefore, the extract from *P. indusiatus* was selected for further testing on its anti-inflammatory effects (Figure 3).

3.3. Optimization of PBMC Culture Conditions Using Whole Blood Samples and FIP Fluid

The characteristics of PBMCs isolated from both whole blood of donor cat and FIP fluid revealed that PBMCs from whole blood contained a small number of platelets and red blood cell contamination (Figure 4A). In contrast, PBMCs isolated from FIP fluid were less pure in terms of white blood cells compared to those from whole blood. The PBMCs sometimes exhibited platelet clumping, which was attributed to the diseased state of the animals. Occasionally, proteins were also present, contributing to the platelet clumping (Figure 4B). Therefore, for the PBMCs isolated from FIP fluid, samples without platelet clumping will be used (Figure 4C).

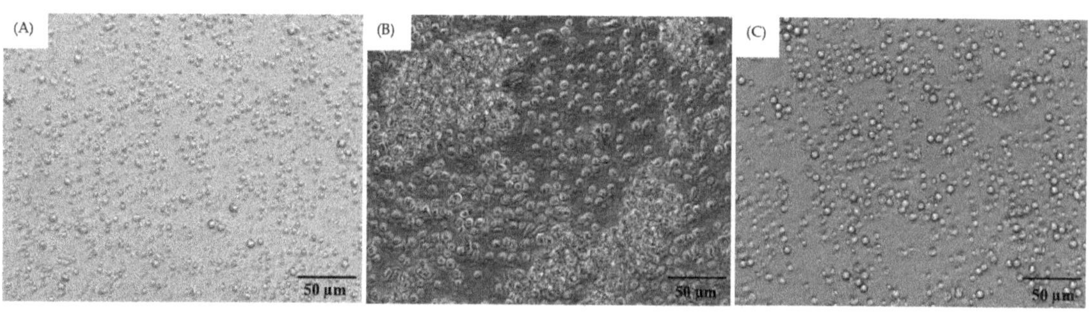

Figure 4. Morphological studies by inverted microscope at actual magnification 50 μm on PBMCs. Normal PBMCs without clumping can be observed when isolated from whole blood (**A**). PBMCs isolated from the fluid showed signs of platelet clumping (**B**). PBMCs separated from FIP fluid without clumping are shown in (**C**).

The experimental results show that at a DMSO concentration of 0.5%, the viability of PBMCs (1×10^6 cells/mL) isolated from whole blood and FIP fluid was 93.75 ± 2.92% and 90.23 ± 2.1%, respectively. This concentration is suitable for further use as a solvent for extracts as it is non-toxic to the cells. However, concentrations of 1%, 1.5%, and 2% were found to be unsuitable as solvents since the viability of PBMCs was below 90%. Therefore, in this study, DMSO at a concentration not exceeding 0.5% was chosen as the solvent for the crude mushroom extract, because PBMCs isolated from both whole blood and FIP fluid had a viability percentage greater than 90% (shown in Figure S1)

The viability assessment of PBMCs (1×10^6 cells/mL) isolated from whole blood and FIP fluid, conducted using the Trypan Blue Exclusion Assay, revealed that on Day 0, cell viability exceeded 90% for both sources, making them suitable for experiments. By Day 1, PBMCs from whole blood remained suitable for experimentation, while PBMCs from FIP fluid were not, as their viability had dropped below 90%. When comparing each day to Day 0, no statistically significant differences were observed from Day 1 to Day 3, whereas statistically significant differences were observed from Day 4 to Day 5.

A comparison of PBMC viability isolated from both whole blood and FIP fluid was performed using a two-way ANOVA ($p < 0.05$). The results indicated no statistically significant differences in PBMC viability between the two sample types from Day 0 to Day 4. However, on Day 5, a statistically significant difference in PBMC viability was observed between PBMCs isolated from whole blood and FIP fluid (Figure 5).

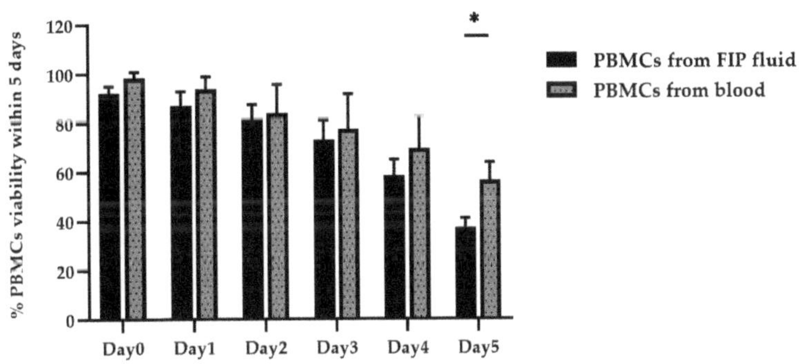

Figure 5. The PBMC viability was analyzed by comparing PBMCs from whole blood collection and FIP fluid samples using a two-way ANOVA. The data are presented as mean ± SEM for PBMCs viability, comparing each cell mean with the other cell means in the same row ($n = 5$). * $p < 0.05$, PBMCs from FIP fluid vs. PBMCs from blood.

However, in this study, PBMCs isolated from samples will be used for testing within 24 h, as they exhibit the highest cell viability, exceeding 90% [22]. This summary confirms that PBMCs isolated from both donor and fluid of FIP retain high viability and are appropriate for subsequent experimental applications.

3.4. LPS-Stimulated PBMCs

PBMCs isolated from donor cat blood samples were stimulated with LPS at concentrations ranging from 1 to 0.001 μg/mL for 24 h. Cell viability and proliferation were then assessed using the CCK-8 colorimetric assay. The results showed that increasing LPS concentrations significantly enhanced cell viability. In the control group without LPS stimulation, cell viability was 91.60 ± 0.85%, whereas at the highest LPS concentration (1 μg/mL), cell viability increased to 99.21 ± 0.10% (Figure 6).

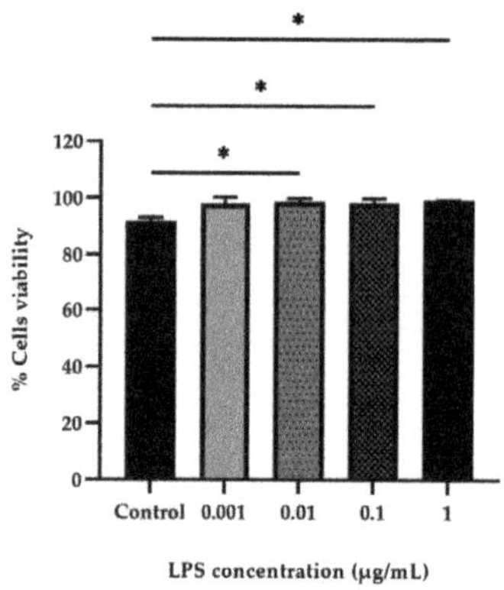

Figure 6. Testing the appropriate concentration of LPS compared to the control group without adding LPS, using concentrations of 0.001, 0.01, 0.1, and 1 μg/mL for 24 h, which were analyzed by one-way ANOVA; * $p < 0.05$ revealed statistically significant differences ($n = 3$).

Subsequently, a comparison of PBMCs viability at each concentration with the control group without LPS addition revealed that at a concentration of 0.001 μg/mL, there was no statistically significant difference compared to the control group. However, at concentrations of 0.01, 0.1, and 1 μg/mL, there were statistically significant differences from the control group (Figure 6). The concentrations of LPS used did not exhibit cytotoxicity, as cell viability exceeded 90% in all cases. Therefore, an LPS concentration of 1 μg/mL was selected to stimulate its effect on inducing cellular inflammation.

3.5. Effect of *P. indusiatus* on the Cell Viability of PBMCs

The toxicity test of *P. indusiatus* extract at various concentrations on PBMCs derived from the whole blood of donor cats over 24 h was assessed using the CCK-8 colorimetric assay. The results revealed that at concentrations of 10,000, 5000, 2500, 625, 156, 78, and 39 μg/mL, the extract remained toxic to the cells, as cell viability was below 90%. The PBMCs exhibited morphological changes, such as reduced size and loss of sheen, indicating cell death, which was distinct from the control group. However, at a concentration of 19 μg/mL, the cells retained their normal morphology, showing no changes compared to the control group, which is consistent with the 90% viability results. Consequently, a concentration of 19 μg/mL is suitable for further testing. The IC_{50} value was calculated to be 1890 μg/mL (as shown in Table S1).

3.6. Evaluation of Nitrite Scavenging Activity of *P. indusiatus* Extract in LPS-Stimulated PBMCs

The evaluation of nitrite (NO_2^-) concentration, resulting from the oxidation of nitric oxide (NO), which serves as an indicator for assessing inflammation in LPS-stimulated cells, revealed statistically significant differences compared to the control group at LPS concentrations of 1, 0.1, 0.01, and 0.001 μg/mL (Figure 7A). The nitrite production measured showed a concentration of 7.7 μM, indicating the induction of inflammation in PBMC stimulated with LPS at the respective concentrations. However, considering the viability of PBMCs, it was found that at a concentration of 1 μg/mL, the cells exhibited the highest viability, consistent with the cell viability results assessed using the CCK-8 assay, which showed 99.21 ± 0.01% (Figure 7A). Therefore, an LPS concentration of 1 μg/mL was selected for inducing inflammation in PBMCs.

The test results showed that PBMCs stimulated with 1 μg/mL LPS (negative control) released a nitrite concentration of 19.73 μM. In contrast, the group treated with the anti-inflammatory drug dexamethasone at 5 μg/mL (positive control) exhibited a reduced nitrite concentration of 12.56 μM. Moreover, the group treated with 19 μg/mL of *P. indusiatus* mushroom extract demonstrated a further reduction in nitrite concentration to 13.36 μM. These results suggest that the *P. indusiatus* mushroom extract may have potential anti-inflammatory properties (Figure 7B).

The results also showed that PBMCs isolated from FIP fluid exhibited inflammation, as assessed by the nitrite concentration shown in Figure 7C, which measured 15.87 μM without stimulation by LPS. This indicates that the cat with FIP already had inflammation in its PBMCs. In the group treated with the anti-inflammatory drug dexamethasone (positive control), the nitrite concentration was 16.13 μM, which was not statistically different from the control group. Similarly, the test group treated with *P. indusiatus* extract had a nitrite concentration of 16.52 μM, which also showed no significant difference compared to the control group.

Figure 7D shows a statistically significant difference between PBMCs isolated from blood and those from FIP fluid. PBMCs from FIP fluid exhibited a higher level of cellular inflammation compared to those from blood. However, when comparing PBMCs isolated from blood and stimulated with LPS to PBMCs from FIP fluid, no significant difference was found (Figure 7D). This statistical analysis confirms that PBMCs isolated from FIP fluid are true inflammatory responses. These results indicate that, since there is no significant difference between PBMCs isolated from blood stimulated with LPS and PBMCs from FIP fluid, the latter are naturally inflamed [23]. In the experimental groups treated with dexamethasone (positive control) and the extract of *P. indusiatus*, a statistically significant difference was observed. PBMCs isolated from blood showed a reduction in nitrite production in both the positive control group and the group treated with *P. indusiatus* extract. However, PBMCs isolated from FIP fluid did not exhibit a reduction in nitrite production.

3.7. Phytochemical Composition and Literature Support for Anti-Inflammatory Activity

In this study, a crude extract of *P. indusiatus* was prepared following the extraction protocol described by Theeraraksakul, K. et al., 2023 [21]. Phytochemical analysis confirmed the presence of key anti-inflammatory constituents, including phenolic compounds, flavonoids, and tannins. The extract demonstrated inhibitory effects on the production of nitric oxide (NO), interleukin-1 (IL-1), interleukin-6 (IL-6), and tumor necrosis factor-alpha (TNF-α). Furthermore, their study reported that the crude extract of *P. indusiatus* also contained various bioactive compounds, such as catechins, flavonoids, polysaccharides, monosaccharides, mucopolysaccharides, allantoin, alkaloids, epicatechin, gallic acid, caffeic acid, and other polyphenolic compounds are also important contributors to the anti-inflammatory activity of the extract [24–26]. Although the crude extract of *P. indusiatus* has not been previously evaluated for its anti-inflammatory effects in cat PBMCs, previous studies have identified bioactive compounds within the extract that possess anti-inflammatory properties. In this study, the observed reduction in NO production in LPS-stimulated PBMCs isolated from healthy cats serves as an important in vitro model (Figure 7B). These findings provide a basis for the future development of anti-inflammatory strategies targeting monocytes and macrophages infected with the FIP virus (Table 1).

Bioactive compounds from *P. indusiatus* demonstrate biochemical and cellular mechanisms of action, ranging from general antioxidant and anti-inflammatory effects to the modulation of specific signaling pathways [7]. Therefore, based on the findings of this study and prior research, there is compelling evidence that extracts from *P. indusiatus* possess anti-inflammatory properties and may modulate the immune response, potentially contributing to the inhibition of FIPV in cats.

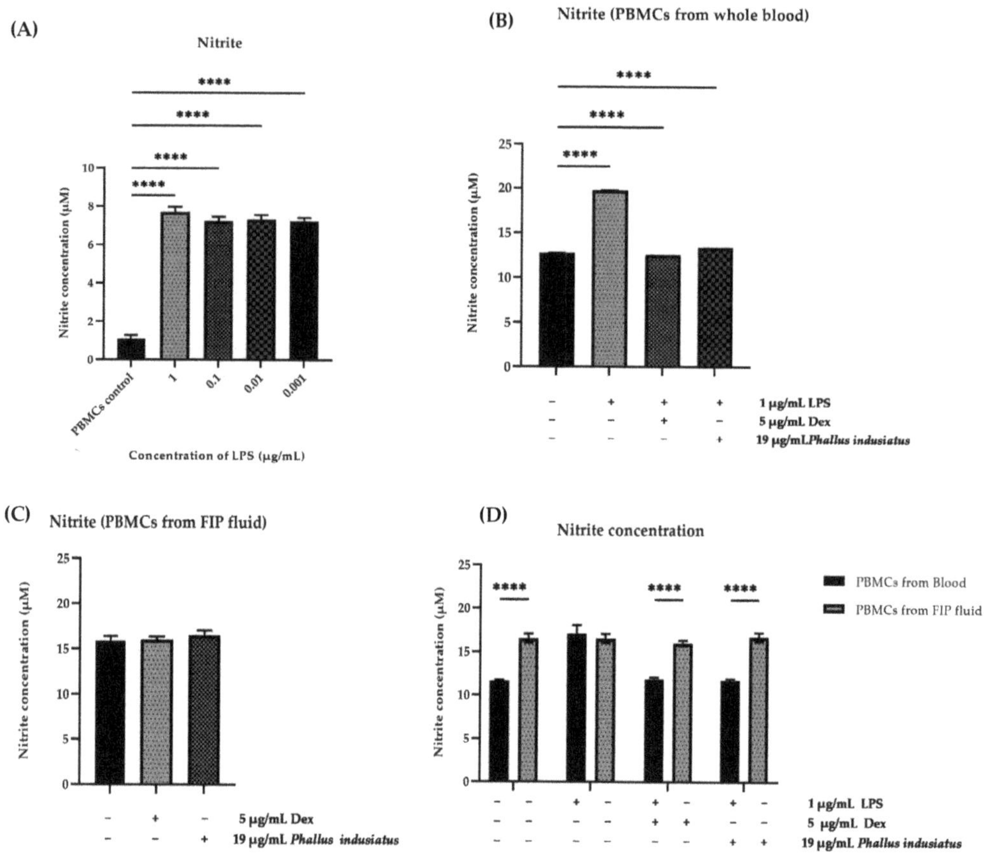

Figure 7. The effect of LPS on nitrite production in stimulated PBMCs was quantified after 24 h of exposure to LPS at concentrations of 0.001, 0.01, 0.1, and 1 µg/mL (**A**). The effect of 19 µg/mL *P. indusiatus* and 5 µg/mL dexamethasone (Dex) on nitrite production in stimulated PBMCs isolated from whole blood of donor cats was quantified after 24 h. The data are presented as mean ± SEM, comparing the mean of each column with the means of every other column (n = 8) (**B**). The stimulation and suppression of inflammation in PBMCs isolated from FIP fluid using *P. indusiatus* extract (n = 5) (**C**). There was analysis by one-way ANOVA; $p < 0.05$ reveals statistically significant differences. The difference in response to treatments between PBMCs isolated from whole blood and those from FIP fluid was analyzed using two-way ANOVA ($p < 0.05$) (n = 4) (**D**). **** $p < 0.0001$.

Table 1. Pharmacological results of *P. indusiatus* efficacy from in vitro studies.

Preparation	Bioactive Compound	Experimental Model	Key Findings	References
Crude extract, hot water, and ethanol.	Phenolics, flavonoids, and tannins.	Murine macrophages used to study the anti-inflammatory effects of LPS stimulation.	The anti-inflammatory activity was demonstrated by a reduction in the level of released nitric oxide (NO).	Theeraraksakul, K. et al., 2023 [21]

Table 1. *Cont.*

Preparation	Bioactive Compound	Experimental Model	Key Findings	References
Crude extract, peel and green mixture, core, and whole mushroom.	Catechins, flavonoids, polysaccharides, monosaccharides, mucopolysaccharides, allantoin, alkaloids, polyphenolic compounds, epicatechin, gallic acid, and caffeic acid.	The experiment was conducted using LPS-induced RAW 264.7 macrophages.	The anti-inflammatory activity was demonstrated by the inhibition of NO, IL-1β, IL-6, and TNF-α secretion.	Ruksiriwanich, W. et al., 2022., Nazir, Y., et al. 2021., Sedtananun, S. et al., 2025 [24–26]
Hot water extract of crude polysaccharide from mushroom fruiting body.	Polysaccharides.	Colitis was induced in mice using dextran sulfate sodium (DSS).	Reduced the expression of inducible iNOS	Kanwal, S. et al., 2020 [27]
This study utilized an ethanolic extract derived from the whole mushroom.	Phenolics, flavonoids, and tannins. [21]	PBMCs from healthy cats were stimulated with LPS and PBMCs from FIP fluid	The anti-inflammatory activity was demonstrated by the inhibition of NO in PBMCs isolated from healthy cats.	Hlaoperm, C. et al.

4. Discussion

In this study, the FIPV M^{pro} had a size of 34 kDa, which increased to 37.2 kDa after the addition of a 10 × His tag and a TEV cleavage site at the N-terminus. The FIPV M^{pro} protein was induced by IPTG and analyzed both quantitatively and qualitatively (Figure 2A). Notably, the FIPV M^{pro} comparable in size to the SARS-CoV-2 M^{pro}, which has a molecular weight of approximately 33 kDa [28]. These findings indicate that the research was successful at this stage.

The present study revealed a significant finding: the mushroom extract demonstrated FIPV protease inhibition at 69.2%, indicating its potential as a natural source of antiviral compounds. This activity is slightly lower than both ritonavir (71.88%) and lopinavir (84.13%), both of which are established broad-spectrum protease inhibitors clinically used in Human Immunodeficiency Virus (HIV) therapy [29] (Figure 3). These agents have also been shown to inhibit SARS-CoV by strongly binding to its M^{pro} [30,31]. In our study, they served as general protease inhibitory benchmarks, since they are not used for treating FIP. It is important to note that this protease inhibition mechanism differs from that of currently used FIP treatments such as molnupiravir or GS-441524, which primarily target the viral RNA-dependent RNA polymerase (RdRp), and are therefore not appropriate controls for a protease inhibition assay [32,33]. The absence of GC376, a highly specific FIPV M^{pro} inhibitor [34,35], was due to its unavailability in our laboratory at the time of the study.

The experimental results revealed that *P. indusiatus* at a concentration of 10 μg/mL achieved the highest inhibition of FIPV M^{pro} activity among the tested mushroom extracts, with an inhibition rate of 69.2% (Figure 3). Despite the mushroom extract's inhibition rate being marginally lower than ritonavir and lopinavir, its efficacy as a natural extract warrants further investigation, suggesting the presence of bioactive molecules with FIPV M^{pro} inhibitory properties. This finding is particularly relevant because, while current gold standard FIP treatments focus on RdRp, viral proteases (specifically M^{pro} or 3CL^{pro}) represent an equally critical and well-validated drug target for coronaviruses due to their essential role in processing polyproteins translated from viral RNA [35–37]. This enzyme is indispensable for viral polyprotein processing, which is a fundamental step for replication [37,38]. The success of protease inhibitors in other viral infections (e.g., Human HIV and Hepatitis C virus [HCV]), along with the promising results of FIPV-specific protease inhibitors such as GC376, underscores the druggability of this enzyme [39]. Also, due to the high conservation of M^{pro} during viral mutagenesis and its critical role in the coronavirus life cycle, M^{pro} is considered as a pivotal target for antivirus drug development [16]

Mushroom extracts are widely recognized for their antiviral and anti-inflammatory properties [40,41]. Previous studies have shown that polysaccharides and adenosine from crude mushroom extract inhibited SARS-CoV protein expression, genomic RNA synthesis, and spike protein activity [42–44]. El-Mekkawy, Sahar et al. (1998) have reported that triterpenes extracted from the fruiting bodies of *Ganoderma lucidum* can inhibit the essential protease enzyme of HIV-1 at a concentration of 7.8 μg/mL [45]. The findings of this study are consistent with previous reports. The crude extract from *P. indusiatus* is particularly noteworthy for further investigation of its anti-inflammatory properties, as it demonstrated the highest enzyme inhibition among the tested mushroom species and also showed potential for inhibiting the FIPV M^{pro}. Exploring protease inhibitors from diverse natural sources, such as mushroom extracts, represents a vital strategy—offering novel compounds, alternative mechanisms to combat resistance, and a foundation for future combination therapies. Therefore, in this study, the crude extract from *P. indusiatus* was selected for testing its anti-inflammatory effects on PBMCs of donor cats and FIP fluid.

The PBMCs isolated from both the whole blood of donor cats and FIP fluid, characterized by monocytes and macrophages, are essential components of the innate immune system [46–48]. The selection of samples should prioritize PBMCs that are already isolated, do not clump, and exhibit a round, glossy appearance (Figure 4). The HIV/AIDS Network Coordination Standard Operating Procedure (HANC-SOP) recommends limiting the processing time to a maximum of 8 h [49]. Delays exceeding 24 h have been associated with decreased cell viability [22]. However, in this study, PBMCs were successfully separated from both whole blood and FIP fluid and tested within 24 h (Figure 5).

In the present study, we employed LPS to activate PBMCs as a model of inflammation. LPS, commonly referred to as endotoxin, is a potent pathogen-associated molecular pattern (PAMP) derived from the outer membrane of Gram-negative bacteria. It is recognized primarily by Toll-like receptor 4 (TLR4) expressed on the surface of immune cells, including monocytes/macrophages [50,51]. Engagement of TLR4 by LPS initiates downstream signaling cascades, notably the myeloid differentiation primary response 88 (MyD88)-dependent pathway, culminating in the activation of key transcription factors such as NF-κB and activator protein 1 (AP-1) [9,50,52,53]. Although FIPV is not a bacterial pathogen, viruses similarly trigger innate immune responses by expressing PAMPs, such as double-stranded RNA (dsRNA) and specific viral proteins [54]. These are recognized by pattern recognition receptors (PRRs), including endosomal Toll-like receptors (TLRs), such as TLR3, TLR7, TLR8, and TLR9, as well as cytosolic sensors like retinoic acid-inducible gene I (RIG-I) and melanoma differentiation-associated protein 5 (MDA5) [55,56]. Activation of these receptors initiates similar signaling pathways that converge on the activation of transcription factors NF-κB and AP-1 [10,11] (Figure 1). Although LPS was used as an inflammatory stimulant in this study, mimicking viral infection in feline PBMCs remains challenging due to the lack of a suitable viral culture system in our lab. Nevertheless, both LPS (via TLR4) and viral components (via other TLRs or cytosolic receptors) activate key pro-inflammatory pathways, particularly NF-κB and AP-1. This results in the upregulation of mediators such as TNF-α, IL-1β, IL-6, and iNOS, supporting the use of LPS as a relevant initial model for evaluating general anti-inflammatory effects.

The innate immune system's ability to initiate responses to LPS was also examined [57]. In this report, 0.001, 0.01, 0.1, and 1 μg/mL of LPS were used to stimulate PBMCs isolated from whole blood of donor cats, compared with the absence of LPS (Figure 6). The data presented in this publication indicate that LPS increased cell viability by approximately 98–99% within 24 h, correlating with the report by Jansky et al. (2003) [58]. A study reported that LPS stimulation increased PBMCs proliferation by activating monocytes and T lymphocytes, thereby promoting their proliferation [59–61]. Based on the study results, LPS may have the potential to increase the number of PBMCs [59–61]. However, PBMCs isolated from FIP fluid do not exhibit LPS-induced inflammatory responses, likely because virus-infected cats already present with systemic leukocyte activation [23]. This diminished responsiveness may be attributed to the fact that FIPV specifically targets monocytes and macrophages, leading to their functional dysregulation [62,63].

The effect of *P. indusiatus* on nitrite scavenging activity in LPS-stimulated PBMCs was assessed. During the inflammatory response, nitric oxide production, measured as nitrite levels in the PBMCs' supernatant, can be upregulated in freshly isolated PBMCs by in vitro stimulation with LPS [64–68]. LPS triggers the TLR4 pathway, leading to the activation of NF-κB and the upregulation of iNOS [50,51]. In LPS-induced samples, monocytes/macrophages are the primary NO producers (Figure 7A) [50,52]. In this study, a group of PBMCs, isolated from donor cat whole blood and stimulated with 1 μg/mL LPS, was evaluated for its anti-inflammatory effects, with a reduction in nitrite serving as an indicator [68]. The crude extract of *P. indusiatus* may possess potential anti-inflammatory properties (Figure 7B).

The test on the reduction of nitrite production by PBMCs isolated from FIP fluid (Figure 7D) clearly demonstrates that cats suffering from this disease naturally exhibit PBMC inflammation, even in the absence of LPS stimulation. The PRRs, such as TLR3, TLR7, RIG-I, and MDA5, lead to IFN signaling [69]. This may trigger IFN responses (IFN-α, IFN-β, IFN-γ), which can either promote NO production or have other immune effects [54]. The host's recognition of viral DNA initiates multiple signal transduction pathways that converge on iNOS upregulation, leading to increased endogenous NO production [70]. The oxidative degradation of NO results in the formation of its breakdown products, nitrate and nitrite. This process, primarily identified in macrophages, involves the expression of the inducible iNOS enzyme [71,72]. Consequently, nitrite levels are significantly higher in PBMCs isolated from FIP fluid compared to those from the whole blood of healthy cats, with a statistically significant difference. This aligns with the research conducted by Özbek et al. (2022), which investigated serum nitric oxide levels in cats affected by FIP [73]. This elevation in nitrite serves as a reliable indicator of inflammation (Figure 7D) [74,75].

The results of this study show a statistically significant difference in the response to treatments between PBMCs isolated from whole blood and those from FIP fluid. PBMCs from whole blood responded to both treatments (dexamethasone and *P. indusiatus* extract) by reducing nitrite production, indicating potential anti-inflammatory effects. However, PBMCs from FIP fluid (possibly from an inflamed environment) did not respond in the same way (Figure 7C,D). The lack of anti-inflammatory effects from both the mushroom extract and dexamethasone in PBMCs from FIP-affected cats underscores the complex and severe nature of FIP-associated inflammation [23,54,62,63,69]. Unlike the acute response induced by LPS, FIP involves chronic, systemic inflammation driven by persistent viral replication in macrophages and a dysregulated immune response, characterized by persistently elevated levels of various inflammatory mediators (a systemic cytokine storm) [52,57,76,77]. In this hyper-inflammatory state, NO production may be maximally upregulated and "locked in", reducing responsiveness to conventional treatments, including dexamethasone [54–56]. This suggests that FIP-related inflammation may involve distinct or more resistant pathways, as well as nitric oxide production through multiple mechanisms, compared to LPS models [10,11,54,56]. Additionally, the direct presence PBMCs from FIP fluid may alter cellular signaling compared to LPS-stimulated healthy PBMCs, potentially affecting drug responsiveness. The previous study suggests that dexamethasone has direct anti-inflammatory effects, acting by altering gene transcription—upregulating anti-inflammatory genes (e.g., IL-10), downregulating pro-inflammatory ones (e.g., IL-6, TNF-α, COX-2, iNOS), and inhibiting the NF-κB pathway [78,79]. In contrast, *P. indusiatus* modulates immunity and oxidative stress through compounds like β-glucans, which stimulate TNF-α, IL-6, and IL-10 to restore immune balance [7,80]. While dexamethasone suppresses NF-κB, *P. indusiatus* fine-tunes it to regulate immune responses [81,82].

Limitations of this study include, firstly, the use of LPS as a model of a general inflammatory stimulus instead of direct FIPV infection. Although both can activate similar inflammatory pathways (e.g., NF-κB leading to TNF-α, IL-1β, IL-6, and iNOS expression). FIPV infection is more complex, involving intricate virus–host interactions that LPS cannot fully mimic. Secondly, attempts to develop an in vitro FIPV infection model were unsuccessful. Although efforts were made to establish an FIPV infection model by infecting CRFK (Crandell-Rees Feline Kidney) cells in vitro, these attempts failed due to challenges in isolating the virus from FIP fluid samples, difficulties in quantifying viral load, and the inconsistency of infection. As a result of these limitations, it was not possible to use the actual virus in the study, making it necessary to rely on the LPS-based simulation system, which may not fully reflect the true nature of FIPV infection.

Future investigations may consider employing alternative approaches to better mimic the natural course of FIPV infection. For example, pseudotyped viral systems expressing the FIPV spike protein could provide a safe and controllable platform for examining viral entry and replication, while FIPV-infected cell lines may offer a more physiologically relevant model for studying virus–host interactions [83,84]. Although these models are not without limitations, they could complement the LPS-based stimulation system used in this study and thereby provide a more comprehensive evaluation of the antiviral and anti-inflammatory potential of the tested extracts.

Despite this, this study represents the first preliminary investigation to evaluate extracts from the mushroom *P. indusiatus* for both anti-FIPV M^{pro} activity and anti-inflammatory effects. It also represents the first assessment of the extract's effects on PBMCs isolated from FIP fluid compared to PBMCs from healthy donors. The extract demonstrated strong inhibitory effects on FIPV M^{pro}. Based on these results, this study is considered successful and has provided important and useful information, opening opportunities for developing new compounds, alternative mechanisms to overcome resistance, and serving as a foundation for future combination therapy research.

Future research should focus on elucidating the roles of pro-inflammatory cytokines (e.g., IL-6, TNF-α) and other relevant inflammatory mediators involved in the pathogenesis of FIP. Bioactive constituents from *P. indusiatus* should be investigated, and their chemical structures elucidated using advanced analytical techniques (e.g., liquid chromatography–mass spectrometry; LC-MS/MS). Subsequent evaluations should include both in vitro and in vivo studies to determine the pharmacological properties of the isolated compounds. Initial in vitro assays should assess antiviral efficacy and anti-inflammatory activity in relevant cell-based models. Promising compounds should then undergo in vivo validation using appropriate animal models. Additionally, comparative analyses with currently available FIPV-specific therapeutics are recommended to benchmark their relative efficacy and therapeutic potential.

5. Conclusions

This study provides compelling preliminary evidence for the dual antiviral and anti-inflammatory potential of *P. indusiatus* extract in the context of FIP. The extract showed a strong inhibitory effect (69.2%) on the FIPV M^{pro}, comparable to established protease inhibitors such as ritonavir and lopinavir. It also significantly reduced NO production in LPS-stimulated PBMCs from healthy cats, indicating potent anti-inflammatory activity. Importantly, the extract was also tested on PBMCs isolated from FIP fluid, which are characterized by chronic inflammation and immune dysregulation. However, these cells showed limited responsiveness (nonresponse) to both dexamethasone and *P. indusiatus* extract, underscoring the complexity of immune activation in FIP. Although LPS was used as a surrogate inflammatory stimulus and a direct FIPV infection model was not available, this is the first study to evaluate *P. indusiatus* for both antiviral and anti-inflammatory effects in the context of FIP. These findings highlight the potential of *P. indusiatus* as a natural source of immunomodulatory agents. Future research should focus on isolating its active compounds, elucidating their mechanisms in relevant infection models, and evaluating their therapeutic potential in vivo, including direct comparisons with FIPV-specific inhibitors.

Supplementary Materials: The following supporting information can be downloaded at https://www.mdpi.com/article/10.3390/vetsci12090847/s1, Figure S1: The appropriate concentration of DMSO for the viability of PBMCs., and Table S1: Optimization of *P. indusiatus* PBMC viability.

Author Contributions: Conceptualization, C.H., A.P., K.C., O.R., and J.R.; methodology, C.H., W.M., E.W., and W.K.; software, C.H., E.W., and W.K.; validation, C.H., A.P., O.R., and J.R.; formal analysis, C.H., E.W., and W.K.; investigation, C.H., E.W., and W.K.; resources, A.R., A.P., K.C., O.R., and J.R.; data curation, C.H.; writing—original draft preparation, C.H.; writing—review and editing, C.H., W.M., E.W., A.P., K.C., O.R., and J.R.; visualization, C.H., E.W., and W.K.; supervision, A.P., K.C., O.R., and J.R.; project administration, J.R.; funding acquisition, A.P., K.C., and J.R. All authors have read and agreed to the published version of the manuscript.

Funding: This research is funded by the High-Quality Research Graduate Development Cooperation Project between Kasetsart University and the National Science and Technology Development Agency (NSTDA), Kasetsart University Research and Development Institute (KURDI), grant number FF(KU) 7.68 and The Faculty of Veterinary Medicine, Kasetsart University, Bangkok, Thailand

Institutional Review Board Statement: This study was approved by the Ethics Committee of Kasetsart University under protocol number ACKU67-VET-124 and all cat owners provided written consent.

Informed Consent Statement: Not applicable.

Data Availability Statement: The raw data supporting the conclusions of this article will be made available by the authors on request. Further inquiries can be directed to the corresponding author.

Acknowledgments: I would like to thank Somruedee Sommart, a veterinary nurse at the Kasetsart University Veterinary Teaching Hospital, for collecting whole blood from donors. I am also grateful to the staff the feline clinic, blood bank unit, emergency room, outpatient department (OPD), and Central Lab at the Kasetsart University Veterinary Teaching Hospital, Faculty of Veterinary Medicine, Kasetsart University, Bangkhen Campus for granting access to the data. Additionally, I extend my gratitude to Nuttanit Jirapanth, a veterinary nurse at the Faculty of Veterinary Medicine, Kasetsart University, for her collaboration with The Nine Animal Hospital and M.J. Animal Hospital in screening and selecting FIP fluid samples.

Conflicts of Interest: The authors confirm that they have no competing interests.

References

1. Hartmann, K. Feline infectious peritonitis. *Vet. Clin. N. Am. Small Anim. Pract.* **2005**, *35*, 39–79. [CrossRef] [PubMed]
2. Decaro, N.; Mari, V.; Lanave, G.; Lorusso, E.; Lucente, M.S.; Desario, C.; Colaianni, M.L.; Elia, G.; Ferringo, F.; Alfano, F.; et al. Mutation analysis of the spike protein in Italian feline infectious peritonitis virus and feline enteric coronavirus sequences. *Res. Vet. Sci.* **2021**, *135*, 15–19. [CrossRef]

3. Gao, Y.Y.; Wang, Q.; Liang, X.Y.; Zhang, S.; Bao, D.; Zhao, H.; Li, S.B.; Wang, K.; Hu, G.X.; Gao, F.S. An updated review of feline coronavirus: Mind the two biotypes. *Virus Res.* **2023**, *326*, 199059. [CrossRef]
4. Xing, L.; Remick, D.G. Relative cytokine and cytokine inhibitor production by mononuclear cells and neutrophils. *Shock* **2003**, *20*, 10–16. [CrossRef]
5. Addie, D.; Belák, S.; Boucraut, B.C.; Egberink, H.; Frymus, T.; Gruffydd, J.T.; Hartmann, K.; Hosie, M.J.; Lloret, A.; Lutz, H.; et al. Feline infectious peritonitis: ABCD guidelines on prevention and management. *J. Feline Med. Surg.* **2009**, *11*, 594–604. [CrossRef]
6. Saylor, S.S.; Coggins, S.; Barker, E.N.; Gunn-Moore, D.; Jeevaratnam, K.; Norris, J.M.; Hughes, D.; Stacey, E.; MacFarlane, L.; O'Brien, C.; et al. Retrospective study and outcome of 307 cats with feline infectious peritonitis treated with legally sourced veterinary compounded preparations of remdesivir and GS-441524 (2020–2022). *J. Feline Med. Surg.* **2023**, *25*, 1098612X231194460. [CrossRef]
7. Habtemariam, S. The chemistry, pharmacology and therapeutic potential of the edible mushroom *Dictyophora indusiata* (Vent ex. Pers.) Fischer (syn. *Phallus indusiatus*). *Biomedicines* **2019**, *7*, 98. [CrossRef]
8. Suleiman, W.B.; Shehata, R.M.; Younis, A.M. In vitro assessment of multipotential therapeutic importance of *Hericium erinaceus* mushroom extracts using different solvents. *Bioresour. Bioprocess.* **2022**, *9*, 1–13. [CrossRef]
9. Luo, R.; Zhang, Z.; Li, Y. An examination of the LPS–TLR4 immune response through the analysis of molecular structures and protein–protein interactions. *Cell Commun. Signal.* **2025**, *23*, 142. [CrossRef]
10. Hennessy, C.; McKernan, D.P. Anti-viral pattern recognition receptors as therapeutic targets. *Cells* **2021**, *10*, 2258. [CrossRef]
11. Yu, L.; Gao, F.; Li, Y.; Su, D.; Han, L.; Li, Y.; Zhang, X.; Feng, Z. Role of pattern recognition receptors in the development of MASLD and potential therapeutic applications. *Biomed. Pharmacother.* **2024**, *175*, 116724. [CrossRef]
12. Lee, M.S.; Kim, Y.J. Signaling pathways downstream of pattern-recognition receptors and their cross talk. *Annu. Rev. Biochem.* **2007**, *76*, 447–480. [CrossRef]
13. Subedi, L.; Lee, J.H.; Yumnam, S.; Ji, E.; Kim, S.Y. Anti-inflammatory effect of sulforaphane on LPS-activated microglia potentially through JNK/AP-1/NF-κB inhibition and Nrf2/HO-1 activation. *Cells* **2019**, *8*, 194. [CrossRef] [PubMed]
14. Vuong, W.; Khan, M.B.; Fischer, C.; Arutyunova, E.; Lamer, T.; Shields, J.; Saffran, H.A.; McKay, R.T.; van Belkum, M.J.; Joyce, M.A.; et al. Feline coronavirus drug inhibits the main protease of SARS-CoV-2 and blocks virus replication. *Nat. Commun.* **2020**, *11*, 4282. [CrossRef] [PubMed]
15. Citarella, A.; Scala, A.; Piperno, A.; Micale, N. SARS-CoV-2 Mpro: A potential target for peptidomimetics and small-molecule inhibitors. *Biomolecules* **2021**, *11*, 607. [CrossRef] [PubMed]
16. Jiang, Z.; Piao, L.; Ren, C.; Zhang, W.; Zhu, Y.; Kong, R. Identifying natural products as feline coronavirus Mpro inhibitors by structure-based virtual screening and enzyme-based assays. *ACS Omega* **2025**, *10*, 2092–2101. [CrossRef]
17. Wang, J.; Wang, F.; Tan, Y.; Chen, X.; Zhao, Q.; Fu, S.; Li, S.; Chen, C.; Yang, H. Crystallization and preliminary crystallographic study of feline infectious peritonitis virus main protease in complex with an inhibitor. *Acta Crystallogr. Sect. F Struct. Biol. Cryst. Commun.* **2014**, *70*, 1612–1615. [CrossRef]
18. Moyadee, W.; Roytrakul, S.; Jaresitthikunchai, J.; Phaonakrop, N.; Choowongkomon, K.; Ploypetch, S.; Tansakul, N.; Rattanasrisomporn, A.; Rattanasrisomporn, J. Serum proteomic approach to identifying differentially expressed proteins in effusive feline infectious peritonitis. *Sci. Rep.* **2025**, *15*, 18899. [CrossRef]
19. Khumtong, K.; Rapichai, W.; Saejung, W.; Khamsingnok, P.; Meecharoen, N.; Ratanabunyong, S.; Dong, H.V.; Tuanthap, S.; Rattanasrisomporn, A.; Choowongkomon, K.; et al. Colorimetric Reverse Transcription Loop-Mediated Isothermal Amplification with Xylenol Orange Targeting Nucleocapsid Gene for Detection of Feline Coronavirus Infection. *Viruses* **2025**, *17*, 418. [CrossRef]
20. Kosukwatthana, P.; Rungsuriyawiboon, O.; Rattanasrisomporn, J.; Kimram, K.; Tansakul, N. Cytotoxicity and Immunomodulatory Effects of Cannabidiol on Canine PBMCs: A Study in LPS-Stimulated and Epileptic Dogs. *Animals* **2024**, *14*, 3683. [CrossRef]
21. Theeraraksakul, K.; Jaengwang, K.; Choowongkomon, K.; Tabtimmai, L. Exploring the biological functions and anti-melanogenesis of *Phallus indusiatus* for mushroom-based cosmetic applications. *Cosmetics* **2023**, *10*, 121. [CrossRef]
22. Bull, M.; Lee, D.; Stucky, J.; Chiu, Y.L.; Rubin, A.; Horton, H.; McElrath, M.J. Defining blood processing parameters for optimal detection of cryopreserved antigen-specific responses for HIV vaccine trials. *J. Immunol. Methods* **2007**, *322*, 57–69. [CrossRef]
23. Bull, M.; Lee, D.; Stucky, J.; Chiu, Y.L.; Rubin, A.; Horton, H.; McElrath, M.J. Study of Macrophage Activity in Cats with FIP and Naturally FCoV-Shedding Healthy Cats. *Pathogens* **2024**, *13*, 437.
24. Nazir, Y.; Linsaenkart, P.; Khantham, C.; Chaitep, T.; Jantrawut, P.; Chittasupho, C.; Rachtanapun, P.; Jantanasakulwong, K.; Phimolsiripol, Y.; Sommano, S.R.; et al. High efficiency in vitro wound healing of *Dictyophora indusiata* extracts via anti-Inflammatory and collagen stimulating (MMP-2 inhibition) mechanisms. *J. Fungi.* **2021**, *7*, 1100. [CrossRef] [PubMed]
25. Ruksiriwanich, W.; Khantham, C.; Linsaenkart, P.; Chaitep, T.; Rachtanapun, P.; Jantanasakulwong, K.; Phimolsiripol, Y.; Režek Jambrak, A.; Nazir, Y.; Yooin, W.; et al. Anti-inflammation of bioactive compounds from ethanolic extracts of edible bamboo mushroom (*Dictyophora indusiata*) as functional health promoting food ingredients. *Int. J. Food Sci. Technol.* **2022**, *57*, 110–122. [CrossRef]

26. Sedtananun, S.; Sujiwattanarat, P.; Hoang Tra, M.H.; Suphakun, P.; Seetaha, S.; Auetragul, A.; Choowongkomon, K.; Tabtimmai, L. *Phallus indusiatus* Extracts Promoted MCF-7 Apoptosis Under TNFα-induced Tumor Microenvironment by Attenuating NF-kappaB and Akt Activation. *Asian Pac. J. Cancer Prev.* **2025**, *26*, 479. [CrossRef]
27. Kanwal, S.; Joseph, T.P.; Aliya, S.; Song, S.; Saleem, M.Z.; Nisar, M.A.; Wang, Y.; Meyiah, A.; Ma, Y.; Xin, Y. Attenuation of DSS induced colitis by *Dictyophora indusiata* polysaccharide (DIP) via modulation of gut microbiota and inflammatory related signaling pathways. *J. Funct. Foods* **2020**, *64*, 103641. [CrossRef]
28. Rong, Y.; Zhang, C.; Gao, W.C.; Zhao, C. Optimization of the expression of the main protease from SARS-CoV-2. *Protein Expr. Purif.* **2023**, *203*, 106208. [CrossRef]
29. Ortiz, R.; Dejesus, E.; Khanlou, H.; Voronin, E.; van Lunzen, J.; Andrade-Villanueva, J.; Fourie, J.; De Meyer, S.; De Pauw, M.; Lefebvre, E.; et al. Efficacy and safety of once-daily darunavir/ritonavir versus lopinavir/ritonavir in treatment-naive HIV-1-infected patients at week 48. *AIDS* **2008**, *22*, 1389–1397. [CrossRef]
30. Schoergenhofer, C.; Jilma, B.; Stimpfl, T.; Karolyi, M.; Zoufaly, A. Pharmacokinetics of lopinavir and ritonavir in patients hospitalized with coronavirus disease 2019 (COVID-19). *Ann. Intern. Med.* **2020**, *173*, 670–672. [CrossRef]
31. Nutho, B.; Mahalapbutr, P.; Hengphasatporn, K.; Pattaranggoon, N.C.; Simanon, N.; Shigeta, Y.; Hannongbua, S.; Rungrotmongkol, T. Why are lopinavir and ritonavir effective against the newly emerged coronavirus 2019? Atomistic insights into the inhibitory mechanisms. *Biochemistry* **2020**, *59*, 1769–1779.
32. Tsika, A.C.; Gallo, A.; Fourkiotis, N.K.; Argyriou, A.I.; Sreeramulu, S.; Löhr, F.; Rogov, V.V.; Richter, C.; Linhard, V.; Gande, S.L.; et al. Binding adaptation of GS-441524 diversifies Macro domains and downregulates SARS-CoV-2 de-MARylation capacity. *J. Mol. Biol.* **2022**, *434*, 167720. [CrossRef]
33. Wang, Z.; Yang, L.; Song, X.Q. Oral GS-441524 derivatives: Next-generation inhibitors of SARS-CoV-2 RNA-dependent RNA polymerase. *Front. Immunol.* **2022**, *13*, 1015355. [CrossRef]
34. Perera, K.D.; Rathnayake, A.D.; Liu, H.; Pedersen, N.C.; Groutas, W.C.; Chang, K.O.; Kim, Y. Characterization of amino acid substitutions in feline coronavirus 3C-like protease from a cat with feline infectious peritonitis treated with a protease inhibitor. *Vet. Microbiol.* **2019**, *237*, 108398. [CrossRef] [PubMed]
35. Pedersen, N.C.; Kim, Y.; Liu, H.; Galasiti Kankanamalage, A.C.; Eckstrand, C.; Groutas, W.C.; Bannasch, M.; Meadows, J.M.; Chang, K.O. Efficacy of a 3C-like protease inhibitor in treating various forms of acquired feline infectious peritonitis. *J. Feline Med. Surg.* **2018**, *20*, 378–392. [CrossRef] [PubMed]
36. Xiong, M.; Su, H.; Zhao, W.; Xie, H.; Shao, Q.; Xu, Y. What coronavirus 3C-like protease tells us: From structure, substrate selectivity, to inhibitor design. *Med. Res. Rev.* **2021**, *41*, 1965–1998. [CrossRef] [PubMed]
37. Zhu, W.; Shyr, Z.; Lo, D.C.; Zheng, W. Viral proteases as targets for coronavirus disease 2019 drug development. *J. Pharmacol. Exp. Ther.* **2021**, *378*, 166–172. [CrossRef]
38. Ablan, O.; Ortega-Alarcon, D.; Jimenez-Alesanco, A.; Ceballos-Laita, L.; Vega, S.; Reyburn, H.T.; Rizzuti, B.; Velazquez-Campoy, A. Structural stability of SARS-CoV-2 3CLpro and identification of quercetin as an inhibitor by experimental screening. *Int. J. Biol. Macromol.* **2020**, *164*, 1693–1703. [CrossRef]
39. Majerová, T.; Konvalinka, J. Viral proteases as therapeutic targets. *Mol. Aspects Med.* **2022**, *88*, 101159. [CrossRef]
40. Varghese, R.; Efferth, T.; Ramamoorthy, S. Exploring the anti-COVID-19 potential of mushroom metabolites: Current status and perspectives. In *Traditional Medicines and Natural Products as Preventive and Therapeutic Agents Against COVID-19*; Academic Press: Cambridge, MA, USA, 2025; pp. 317–337.
41. Chun, S.; Gopal, J.; Muthu, M. Antioxidant activity of mushroom extracts/polysaccharides—Their antiviral properties and plausible anti-COVID-19 properties. *Antioxidants* **2021**, *10*, 1899. [CrossRef]
42. Zhang, Y.; Zhang, G.; Ling, J. Medicinal fungi with antiviral effect. *Molecules* **2022**, *27*, 4457. [CrossRef] [PubMed]
43. Verma, A.K. Cordycepin: A bioactive metabolite of Cordyceps militaris and polyadenylation inhibitor with therapeutic potential against Covid-19. *J. Biomol. Struct. Dyn.* **2020**, *40*, 3745–3752. [CrossRef] [PubMed]
44. Zhao, C.; Gao, L.; Wang, C.; Liu, B.; Jin, Y.; Xing, Z. Structural characterization and antiviral activity of a novel heteropolysaccharide isolated from Grifola frondosa against enterovirus 71. *Carbohydr Polym.* **2016**, *144*, 382–389. [CrossRef] [PubMed]
45. El-Mekkawy, S.; Meselhy, M.R.; Nakamura, N.; Tezuka, Y.; Hattori, M.; Kakiuchi, N.; Shimotohno, K.; Kawahata, T.; Otake, T. Anti-HIV-1 and anti-HIV-1-protease substances from Ganoderma lucidum. *Phytochemistry* **1998**, *49*, 1651–1657. [CrossRef]
46. Bienzle, D.; Reggeti, F.; Clark, M.E.; Chow, C. Immunophenotype and functional properties of feline dendritic cells derived from blood and bone marrow. *Vet. Immunol. Immunopathol.* **2003**, *96*, 19–30. [CrossRef]
47. Parihar, A.; Eubank, T.D.; Doseff, A.I. Monocytes and macrophages regulate immunity through dynamic networks of survival and cell death. *J. Innate Immun.* **2010**, *2*, 204–215. [CrossRef]
48. Lopes, R.; Sampaio, F.; Carvalho, H.L.; Garcês, A.; Fernandes, C.; Neves, C.V.; Brito, A.S.; Marques, T.; Sousa, C.; Silva, A.R.; et al. Feline Infectious Peritonitis Effusion Index: A Novel Diagnostic Method and Validation of Flow Cytometry-Based Delta Total Nucleated Cells Analysis on the Sysmex XN-1000V®. *Vet. Sci.* **2024**, *11*, 563. [CrossRef]

49. Browne, D.J.; Miller, C.M.; Doolan, D.L. Technical pitfalls when collecting, cryopreserving, thawing, and stimulating human T-cells. *Front. Immunol.* **2024**, *15*, 1382192. [CrossRef]
50. Xu, Y.; Jagannath, C.; Liu, X.D.; Sharafkhaneh, A.; Kolodziejska, K.E.; Eissa, N.T. Toll-like receptor 4 is a sensor for autophagy associated with innate immunity. *Immunity* **2007**, *27*, 135–144. [CrossRef]
51. Oshio, T.; Kawashima, R.; Kawamura, Y.I.; Hagiwara, T.; Mizutani, N.; Okada, T.; Otsubo, T.; Inagaki-Ohara, K.; Matsukawa, A.; Haga, T.; et al. Chemokine receptor CCR8 is required for lipopolysaccharide-triggered cytokine production in mouse peritoneal macrophages. *PLoS ONE* **2014**, *9*, e94445. [CrossRef]
52. Tucureanu, M.M.; Rebleanu, D.; Constantinescu, C.A.; Deleanu, M.; Voicu, G.; Butoi, E.; Calin, M.; Manduteanu, I. Lipopolysaccharide-induced inflammation in monocytes/macrophages is blocked by liposomal delivery of Gi-protein inhibitor. *Int. J. Nanomed.* **2018**, *13*, 63–76. [CrossRef]
53. Chen, H.; Wang, F.; Mao, H.; Yan, X. Degraded λ-carrageenan activates NF-κB and AP-1 pathways in macrophages and enhances LPS-induced TNF-α secretion through AP-1. *Biochim. Biophys. Acta* **2014**, *1840*, 2162–2170. [CrossRef]
54. Thompson, M.R.; Kaminski, J.J.; Kurt-Jones, E.A.; Fitzgerald, K.A. Pattern recognition receptors and the innate immune response to viral infection. *Viruses* **2011**, *3*, 920. [CrossRef]
55. Kolli, D.; Velayutham, T.S.; Casola, A. Host-viral interactions: Role of pattern recognition receptors (PRRs) in human pneumovirus infections. *Pathogens* **2013**, *2*, 232–263. [CrossRef] [PubMed]
56. Okamoto, M.; Tsukamoto, H.; Kouwaki, T.; Seya, T.; Oshiumi, H. Recognition of viral RNA by pattern recognition receptors in the induction of innate immunity and excessive inflammation during respiratory viral infections. *Viral Immunol.* **2017**, *30*, 408–420. [CrossRef] [PubMed]
57. Ngkelo, A.; Meja, K.; Yeadon, M.; Adcock, I.; Kirkham, P.A. LPS induced inflammatory responses in human peripheral blood mononuclear cells is mediated through NOX4 and G i α dependent PI-3kinase signalling. *J. Inflamm.* **2012**, *9*, 1–7. [CrossRef]
58. Jansky, L.; Reymanova, P.; Kopecky, J. Dynamics of cytokine production in human peripheral blood mononuclear cells stimulated by LPS, or infected by Borrelia. *Physiol. Res.* **2003**, *52*, 593–598. [CrossRef]
59. Lin, Z.; Huang, Y.; Jiang, H.; Zhang, D.; Yang, Y.; Geng, X.; Li, Y. Functional differences and similarities in activated peripheral blood mononuclear cells by lipopolysaccharide or phytohemagglutinin stimulation between human and cynomolgus monkeys. *Ann. Transl. Med.* **2021**, *9*, 257. [CrossRef]
60. Kubin, M.; Kamoun, M.; Trinchieri, G. Interleukin 12 synergizes with B7/CD28 interaction in inducing efficient proliferation and cytokine production of human T cells. *J. Exp. Med.* **1994**, *180*, 211–222. [CrossRef]
61. Ulmer, A.J.; Flad, H.D.; Rietschel, T.; Mattern, T. Induction of proliferation and cytokine production in human T lymphocytes by lipopolysaccharide (LPS). *Toxicology* **2000**, *152*, 37–45. [CrossRef]
62. Malbon, A.J.; Michalopoulou, E.; Meli, M.L.; Barker, E.N.; Tasker, S.; Baptiste, K.; Kipar, A. Colony stimulating factors in early feline infectious peritonitis virus infection of monocytes and in end stage feline infectious peritonitis; a combined in vivo and in vitro approach. *Pathogens* **2020**, *9*, 893. [CrossRef]
63. Cornelissen, E.; Dewerchin, H.L.; Van Hamme, E.; Nauwynck, H.J. Absence of surface expression of feline infectious peritonitis virus (FIPV) antigens on infected cells isolated from cats with FIP. *Vet. Microbiol.* **2007**, *121*, 131–137. [CrossRef] [PubMed]
64. Hirano, T.; Matsuda, T.; Turner, M.; Miyasaka, N.; Buchan, G.; Tang, B.; Sato, K.; Shimi, M.; Maid, R.; Feldmann, M.; et al. Excessive production of interleukin 6/B cell stimulatory factor-2 in rheumatoid arthritis. *Eur. J. Immunol.* **1988**, *18*, 1797–1802. [CrossRef] [PubMed]
65. Bergroth, V.; Zvaifler, N.J.; Firestein, G.S. Cytokines in chronic inflammatory arthritis. III. Rheumatoid arthritis monocytes are not unusually sensitive to γ-interferon, but have defective γ-interferon–mediated HLA–DQ and HLA–DR induction. *Arthritis Rheum.* **1989**, *32*, 1074–1079. [CrossRef] [PubMed]
66. Firestein, G.S.; Alvaro-Gracia, J.M.; Maki, R. Quantitative analysis of cytokine gene expression in rheumatoid arthritis. *J. Immunol.* **1990**, *144*, 3347–3353. [CrossRef]
67. Sakurai, H.; Kohsaka, H.; Liu, M.F.; Higashiyama, H.; Hirata, Y.; Kanno, K.; Saito, I.; Miyasaka, N. Nitric oxide production and inducible nitric oxide synthase expression in inflammatory arthritides. *J. Clin. Investig.* **1995**, *96*, 2357–2363. [CrossRef]
68. Davies, C.A.; Rocks, S.A.; O'Shaughnessy, M.C.; Perrett, D.; Winyard, P.G. Analysis of nitrite and nitrate in the study of inflammation. In *Inflammation Protocols, Methods in Molecular Biology*; Humana Press: Totowa, NJ, USA, 2003; pp. 305–320.
69. Capozza, P.; Pratelli, A.; Camero, M.; Lanave, G.; Greco, G.; Pellegrini, F.; Tempesta, M. Feline coronavirus and alpha-herpesvirus infections: Innate immune response and immune escape mechanisms. *Animals* **2021**, *11*, 3548. [CrossRef]
70. Garren, M.R.; Ashcraft, M.; Qian, Y.; Douglass, M.; Brisbois, E.J.; Handa, H. Nitric oxide and viral infection: Recent developments in antiviral therapies and platforms. *Appl. Mater. Today* **2021**, *22*, 100887. [CrossRef]
71. Omar, M.; Abdelal, H.O. Nitric oxide in parasitic infections: A friend or foe? *J. Parasit. Dis.* **2022**, *46*, 1147–1163. [CrossRef]
72. Olyslaegers, D.A.; Dedeurwaerder, A.; Desmarets, L.M.; Vermeulen, B.L.; Dewerchin, H.L.; Nauwynck, H.J. Altered expression of adhesion molecules on peripheral blood leukocytes in feline infectious peritonitis. *Vet. Microbiol.* **2013**, *166*, 438–449. [CrossRef]

73. Özbek, M.; Özkan, C.; Kaya, A.; Yıldırım, S.; Kozat, S.; Akgül, Y. Clinicopathological and biochemical evaluation of Feline Infectious Peritonitis in Turkish Van cats. *J. Hell. Vet. Med. Soc.* **2022**, *73*, 4379–4388. [CrossRef]
74. Waltz, P.; Escobar, D.; Botero, A.M.; Zuckerbraun, B.S. Nitrate/nitrite as critical mediators to limit oxidative injury and inflammation. *Antioxid. Redox Signal.* **2015**, *23*, 328–339. [CrossRef] [PubMed]
75. Gao, X.Q.; Fei, F.; Huo, H.H.; Huang, B.; Meng, X.S.; Zhang, T.; Liu, B.L. Impact of nitrite exposure on plasma biochemical parameters and immune-related responses in Takifugu rubripes. *Aquat. Toxicol.* **2020**, *218*, 105362. [CrossRef] [PubMed]
76. Sherding, R.G. Feline infectious peritonitis (feline coronavirus). *J. Small Anim. Pract.* **2009**, *50*, 132–143.
77. Malbon, A.J.; Fonfara, S.; Meli, M.L.; Hahn, S.; Egberink, H.; Kipar, A. Feline infectious peritonitis as a systemic inflammatory disease: Contribution of liver and heart to the pathogenesis. *Viruses* **2019**, *11*, 1144. [CrossRef]
78. Zappia, C.D.; Torralba-Agu, V.; Echeverria, E.; Fitzsimons, C.P.; Fernández, N.; Monczor, F. Antihistamines potentiate dexamethasone anti-inflammatory effects. Impact on glucocorticoid receptor-mediated expression of inflammation-related genes. *Cells* **2021**, *10*, 3026.
79. Jiang, Z.; Zhu, L. Update on molecular mechanisms of corticosteroid resistance in chronic obstructive pulmonary disease. *Pulm. Pharmacol. Ther.* **2016**, *37*, 1–8. [CrossRef]
80. Yadav, D.; Negi, P.S. Role of mushroom polysaccharides in improving gut health and associated diseases. In *Microbiome, Immunity, Digestive Health and Nutrition*; Academic Press: London, UK; Elsevier: Amsterdam, The Netherlands, 2022; pp. 431–448.
81. Thrikawala, S.U.; Anderson, M.H.; Rosowski, E.E. Glucocorticoids Suppress NF-κB–Mediated Neutrophil Control of Aspergillus fumigatus Hyphal Growth. *J. Immunol.* **2024**, *213*, 971–987. [CrossRef]
82. Kumar, Y.; Xu, B. New Insights into Chemical Profiles and Health-Promoting Effects of Edible Mushroom *Dictyophora indusiata* (Vent ex. Pers.) Fischer: A Review. *J. Fungi.* **2025**, *11*, 75. [CrossRef]
83. Liu, S.; Zhang, L.; Fu, W.; Liang, Z.; Yu, Y.; Li, T.; Tong, J.; Liu, F.; Nie, J.; Lu, Q.; et al. Optimization and validation of a virus-like particle pseudotyped virus neutralization assay for SARS-CoV-2. *MedComm* **2024**, *5*, e615. [CrossRef]
84. Luo, H.; Lv, L.; Yi, J.; Zhou, Y.; Liu, C. Establishment of replication deficient vesicular stomatitis virus for studies of PEDV spike-mediated cell entry and its inhibition. *Microorganisms* **2023**, *11*, 2075. [CrossRef]

veterinary sciences

MDPI

Case Report

Thoracic Epidural Anesthesia in Cats: A Retrospective Case Series

Elena Lardone *, Alessandra Landi, Greta Martinelli and Paolo Franci

Department of Veterinary Sciences, University of Turin, 2 Largo Paolo Braccini, 10095 Grugliasco, Italy; alessandra.landi@unito.it (A.L.); greta.martinelli@unito.it (G.M.); paolo.franci@unito.it (P.F.)
* Correspondence: elena.lardone@unito.it; Tel.: +39-(0)116708722

Simple Summary: Cats are highly sensitive to pain-induced stress, which often is displayed through behavioral changes such as anorexia and aggression. Perioperative pain management is a very critical component for a cat as a surgical patient. Despite its well-documented benefits in human medicine, thoracic epidural anesthesia (TEA) remains unexplored in cats due to anatomical challenges and demands for technical expertise. This retrospective case study aimed to describe the feasibility, safety, and analgesic efficacy of TEA in cats undergoing major surgeries. Nine cats received 0.2 mL/kg of 0.5% ropivacaine combined with 0.1 mg/kg morphine via TEA at the T12–T13 level. Thoracic epidural anesthesia was associated with stable intraoperative cardiovascular parameters and reduced systemic perioperative rescue analgesia in the majority of cases. Only one cat exhibited inadequate analgesic coverage, likely due to technical failure. No significant adverse effects or complications were observed. These findings support the potential of TEA as a viable analgesic strategy in surgical cats, with implications for improving clinical outcomes, enhancing animal welfare, and advancing the standard of care in veterinary anesthesia.

Abstract: Thoracic epidural anesthesia (TEA) is widely used in human medicine to provide effective perioperative analgesia, yet its application in veterinary species—particularly cats—remains underexplored. This retrospective case study describes the use of TEA in nine client-owned cats undergoing major surgeries. All cats received a single epidural injection of 0.2 mL/kg of 0.5% ropivacaine combined with 0.1 mg/kg morphine at the T12–T13 interspace using a 25 G × 25 mm Quincke needle. Intraoperative physiological parameters were continuously monitored, and postoperative analgesia was assessed using a validated pain scale. Only one cat exhibited inadequate analgesic coverage, likely due to TEA failure. Of the nine cats, seven required minimal to no intraoperative rescue analgesia, while five received postoperative opioids on the day following surgery. Hemodynamic stability was observed in most cases, with no significant complications reported. These findings suggest that TEA is a feasible and effective technique for perioperative pain management in cats undergoing major surgery. Further prospective studies are warranted to confirm these initial findings and investigate the safety of the technique in a larger population.

Keywords: epidural injection; TEA; ropivacaine; morphine; major surgery

1. Introduction

Thoracic epidural anesthesia (TEA) is widely considered the gold standard for perioperative analgesia in people, particularly in the context of cardiac, thoracic, and cranial abdominal surgical procedures [1]. An extensive body of clinical and experimental research has consistently demonstrated the multifactorial efficacy of TEA. Notably, TEA produces a profound central neural blockade that effectively attenuates the neuroendocrine stress response associated with surgical trauma, thereby mitigating the incidence of perioperative complications [2]. Furthermore, the sympathectomy induced by TEA confers protective effects on vital organ systems, including the cardiovascular, pulmonary, and gastrointestinal systems, while also exerting favorable influences on immune modulation and coagulation dynamics [3]. Consequently, TEA is regarded as a critical factor in enhancing positive postoperative outcomes following major surgical interventions [4].

In veterinary medicine, anatomical limitations in dogs and cats pose technical challenges for epidural needle placement at the thoracic level and TEA remains a relatively under-investigated technique, with limited data available in dogs [5–7]. Tonge et al. [7] reported successful use of TEA in a dog undergoing thoracotomy and pulmonary lobectomy, achieving favorable anesthetic and analgesic outcomes throughout the perioperative period. Similarly, Lardone and colleagues [6] demonstrated the effectiveness of TEA for intra- and postoperative pain management during major thoracic and cranial abdominal procedures. These findings suggest that TEA may offer substantial benefits in veterinary anesthesia, mirroring its established role in human medicine.

To the authors' knowledge, there is currently no published evidence evaluating the application of TEA in cats. Feline patients are generally more susceptible to stress than dogs, and postoperative pain is a key factor capable of eliciting dysfunction across multiple organ systems. In addition, pain and stress can significantly influence behavioral patterns during hospitalization, frequently manifesting as anorexia, inhibition, and aggression [8]. These considerations highlight the critical importance of comprehensive and proactive pain management strategies in cats undergoing major surgical procedures.

In light of the routine implementation of TEA in both dogs and cats at our institution, the present study aims to retrospectively describe our clinical experience with its use in a cohort of cats undergoing a variety of major surgical interventions.

2. Case Presentation

Cats that underwent major surgical procedures between December 2021 and March 2025 at the Teaching Veterinary Hospital of the University of Turin (Grugliasco, Italy) were included in this study. Written informed consent was obtained from all owners. Cases were consecutively numbered from one to nine. Demographic data and relevant clinical details are summarized in Table 1.

2.1. Treatment

Table 1 shows the anesthesia protocol.

Following induction of general anesthesia, endotracheal intubation was achieved using appropriately sized cuffed tubes. The endotracheal tubes were connected to a circular breathing system, delivering a gas mixture of oxygen and air with a fractional inspired oxygen (FiO_2) of 0.40. Maintenance of anesthesia was achieved in three cats using sevoflurane and in six cats using isoflurane. The end-tidal halogenated anesthetic concentration (FE'AA) was set at the anesthetist's discretion to maintain the surgical plane of anesthesia.

All animals underwent volume-controlled intermittent positive pressure ventilation (Primus, Draeger, Germany) to maintain normocapnia, targeting end-tidal carbon dioxide ($PE'CO_2$) levels between 35 and 45 mmHg (4.6 and 6 kPa).

Table 1. Patient and medical details, anesthetic protocol, and intraoperative and postoperative information. DSH: domestic short hair; M: male; F: female; n: neutered; y: years; m: months; min: minutes; IM: intramuscular; IV: intravenous; dex: dexmedetomidine; CRI: constant rate infusion; RA: rescue analgesia; h: hours.

Case	Signalment	Medical State	Surgical Procedure and Anesthesia Time	Premedication	Anesthesia Protocol	Intraoperative RA	Intraoperative Events	First Postoperative RA
Case 1	DSH Mn 7 y 4 kg	Chylothorax	Thoracotomy at the 10th intercostal space and pericardiectomy 200 min	None	Fentanyl (6 μg/kg IV) + Propofol (8 mg/kg IV) Isoflurane Cisatracurium (0.2 mg/kg IV at the beginning of the surgery)	Fentanyl (1 bolus at 2 μg/kg IV)	None	Methadone (0.2 mg/kg IM) 9 h after the end of surgery
Case 2	DSH M 3 m 1 kg	Diaphragmatic hernia Polypnea Weakness	Diaphragmatic herniorrhaphy 50 min	None	Propofol (5 mg/kg IV) + Cisatracurium (0.3 mg/kg IV, as co-induction agent) Sevoflurane	None	None	Methadone (0.4 mg/kg IM) the day after surgery
Case 3	DSH Fn 5 y 3.4 kg	Thymoma	Sternotomy 240 min	Alfaxalone (1 mg/kg IM) Methadone (0.3 mg/kg IM) Dex (4 μg/kg IM)	Alfaxalone (2 mg/kg IV) Sevoflurane	Fentanyl (2 boluses at 2 μg/kg IV + CRI at 5 μg/kg/h IV) Ketamine (1 bolus at 0.5 mg/kg IV + CRI at 20 μg/kg/h IV)	Tachycardia Hypertension Hypotension Noradrenaline	Methadone (0.2 mg/kg IM) at recovery from anesthesia
Case 4	DSH Mn 1 y 4.6 kg	Diaphragmatic hernia Polypnea	Diaphragmatic herniorrhaphy 135 min	Methadone (0.1 mg/kg IV)	Dex (2 μg/kg IV) + Alfaxalone (1 mg/kg IV) Isoflurane	None	None	Methadone (0.2 mg/kg IM) the day after surgery
Case 5	DSH Fn 1 y 2 kg	Fourth aortic duct persistency	Left lateral thoracotomy at the fourth intercostal space 180 min	Methadone (0.1 mg/kg IV)	Dex (2 μg/kg IV) + Propofol (2.5 mg/kg IV) Sevoflurane	Fentanyl (1 bolus at 2 μg/kg IV)	Tachycardia Hypertension Hypotension Bradycardia Atropine	Lack of data
Case 6	DSH Fn 5 y 3.8 kg	Chylothorax	Left lateral thoracotomy at the fourth intercostal space 210 min	Alfaxalone (1 mg/kg IM) Methadone (0.3 mg/kg IM) Dex (2 μg/kg IM)	Propofol 2 mg/kg Isoflurane Cisatracurium (0.2 mg/kg at the beginning of the surgery)	Fentanyl (1 bolus at 2 μg/kg IV) Ketamine (1 bolus at 0.5 mg/kg IV)	Tachycardia Hypertension	Lack of data
Case 7	DSH Fn 10 y 4 kg	Hyperthyroidism Chronic renal failure Aggressive Behavior	Left-side radical mastectomy 70 min	Alfaxalone (1.5 mg/kg IM) Methadone (0.3 mg/kg IM) Midazolam (0.25 mg/kg IM)	Propofol (2 mg/kg IV) Isoflurane	None	None	Methadone (0.2 mg/kg IM) the day after surgery
Case 8	British cat Mn 8 m 4 kg	Peritoneo-pericardial hernia	Herniorrhaphy 120 min	Alfaxalone (1 mg/kg IM) Methadone (0.15 mg/kg IM)	Propofol (3.5 mg/kg IV) Isoflurane	None	None	Methadone (0.2 mg/kg IM) the day after surgery
Case 9	DSH Mn 9 y 5 kg	Thymoma	Sternotomy 240 min	Alfaxalone (1 mg/kg IM) Methadone (0.2 mg/kg IM) Dex (3 μg/kg IM)	Propofol (1 mg/kg IV) Isoflurane	None	None	Methadone (0.2 mg/kg IM) the day after surgery

Cefazolin (Cefazolina 1 g, TEVA, Milan (MI), Italy) was administered intravenously at a dose of 20 mg/kg within 20 min prior to the initial surgical incision. Lactated Ringer's solution (Ringer lattato, Fresenius Kabi, Isolandella Scala (VR), Italy) was infused at a rate of 5 mL/kg/h throughout the anesthetic period. Upon confirmation of a stable surgical plane of anesthesia, cats were placed in lateral recumbency, and the thoraco-lumbar area of the vertebral column was clipped and aseptically prepared with chlorhexidine (4%) and alcohol (70%).

2.2. Monitoring

During the intraoperative period, a multiparametric monitor (Infinity Delta, Draeger, Lubeck, Germany) was employed to continuously assess cardiovascular variables—including invasive and/or non-invasive systolic, diastolic, and mean arterial pressures (SAP, DAP, and MAP), heart rate, and cardiac rhythm—as well as respiratory parameters, including $PE'CO_2$, peak inspiratory pressure, respiratory rate, tidal volume, minute ventilation, FiO_2, and FE'AA. Esophageal temperature was also monitored.

To maintain normothermia (body temperature > 36 °C), an active warming system (Bair Hugger Warmer Model 505, Augustine Biomedical Design, Cape Town, South Africa) was used throughout the perioperative period. All monitored parameters were manually recorded at 5 min intervals for the duration of anesthesia.

2.3. Thoracic Epidural Anesthesia

Cats were positioned in lateral recumbency with slight flexion of the vertebral column, and the dorsal aspect of the body was extended beyond the edge of the surgical table. The spinous and articular processes of the T13 vertebra were identified using the index finger and thumb, respectively. A 25 gauge × 25 mm Quincke spinal needle (Becton, Dickinson and Company, Franklin Lakes, NJ, USA) was introduced on the dependent side at the T12–T13 intervertebral space via a paramedian approach with cephalad angulation, as previously described in humans [9], dogs [10], and cats [11]. Upon encountering the lamina, it was slightly withdrawn and redirected in a ventrocranial–medial direction until resistance from the ligamentum flavum was pierced and perceived. A plain solution of ropivacaine (5 mg/mL; Ropivacaina, Galenica Senese, Monteroni d'Arbia (SI), Italy) at a dose of 1 mg/kg, combined with morphine (10 mg/mL; Morfina Cloridrato, Monico, Venice (VE), Italy) at 0.1 mg/kg, was administered slowly.

2.4. Intraoperative Analgesia

The anesthetists were free to use fentanyl (Fentadon 0.05 mg/mL, Dechra, Turin (TO), Italy) and ketamine (Ketavet 100 mg/mL, MSD Animal Health, Rahway, NJ, USA) as needed to stabilize the anesthetic plane. The amounts of analgesics administered during the procedure are shown in Table 1.

2.5. Recovery and Postoperative Plan

The urinary bladder was voided manually at the end of the surgery. Meloxicam (Metacam 2 mg/mL, Boehringer Ingelheim Animal Health, Milan (MI), Italy) at 0.3 mg/kg subcutaneously was administered when normothermia was confirmed during recovery with the exception of case 2 that did not receive any anti-inflammatory drugs.

Methadone (Semfortan 10 mg/mL, Dechra, Northwich, UK) was given intramuscularly (0.2–0.4 mg/kg, IM) during the postoperative period as rescue analgesia using the Glasgow Composite Measure Pain Scale short form (intervention level of 5/20 and above) [12]. In the majority of cases, time-point assessments of analgesia were conducted at 1, 3, and 5 h (h) following extubation. Five cats received methadone the day after the surgery and a patient was administered 9 h after the end of anesthesia. Case 3 received methadone at recovery from anesthesia. Postoperative data were incomplete for two cats.

The assessment of motor function was predominantly performed at 1, 3, and 5 h after extubation, using a previously published scoring system [13]. At 3 h and 5 h after extubation, the 5/7 cats and the 2/7 cats respectively showed motor deficit (grade 2 and 3). Complete motor function recovery occurred in all cats within 24 h of surgery.

3. Discussion

To the authors' knowledge, this is the first report to describe the clinical application of TEA in cats.

In this case series, the use of TEA along with general anesthesia contributed to reducing the perioperative use of rescue analgesia in some cats. It is noteworthy that seven cats out of nine did not need intraoperative rescue analgesia or received just a fentanyl dose. Given that several studies demonstrated a correlation between reduced intraoperative rescue analgesia and successful regional nerve blockade [14–16], TEA might have provided effective nociceptive control in the present case series.

No reliable conclusions can be drawn regarding the duration or extent of analgesia provided by TEA during the postoperative period. Nonetheless, five cats received methadone the day after surgery and one cat was administered rescue analgesia 9 h after anesthesia. Considering the invasiveness of the procedures performed, this finding may suggest satisfactory analgesic coverage of the single shot TEA in the first 24 h postoperatively.

The epidural injection of local anesthetics results in the blockade of the paravertebral sympathetic chain. While this can occasionally lead to hypotension due to decreased peripheral vasoconstriction, it may also attenuate the sympathetic response to surgical manipulation, thereby enhancing hemodynamic stability [17–19]. The hemodynamic stability and minimal intraoperative rescue analgesia (Table 1), which characterized the majority of the cases, might be attributable to TEA.

Case 3 likely represented a failure of the technique (i.e., incorrect needle positioning or suboptimal anesthetic spread), as it was characterized by hemodynamic instability, insufficient intraoperative antinociceptive effect, and inadequate postoperative analgesic coverage. In this case series, the detection of the epidural space exclusively relied upon the loss of resistance (LOR) at piercing the yellow ligamentum. In this perspective, a Tuohy needle offers the characteristic "pop sensation" and allows the use of air or saline for the LOR technique, which helps confirm the epidural space. Due to its very small diameter, the 25 G Quincke needle made it more difficult to rely upon the LOR technique or other methods for verifying the needle position such as the observation of epidural pressure waves.

In this study, the Quincke spinal needle was preferred over the more commonly used Tuohy needle for epidural anesthesia. The 2.5 cm × 25 G Quincke needle is more suitable for TEA in cats, as it requires minimal needle advancement (about 1 cm [11]). On the other hand, the pediatric 5 cm × 22 G Tuohy needle available would be too long (8 cm long, including the hub). The authors therefore consider the Quincke needle safer than the Tuohy needle due to its shorter length and its greater stability after positioning. However, its smaller gauge makes detecting cerebrospinal fluid reflux more challenging, which can be mitigated by using the dependent paramedian approach. The technique requires more skill, but once mastered, it can be performed quickly and safely. While epidural anesthesia in cats offers notable potential benefits, it also carries a risk of spinal cord injury due to the relatively caudal start of the cauda equina, which provides less space for safe needle placement compared to other species [20]. Given the limited number of cases in this study, it was not possible to assess the risks associated with the technique. Additionally, the inability to use the loss-of-resistance method and the Tuohy needle may further increase the likelihood of spinal cord puncture. For these reasons, TEA should be reserved for specialists with appropriate practical training, a thorough understanding of neuraxial techniques, and the ability to manage patients perioperatively to optimize outcomes and minimize complications.

Although the benefits of continuous epidural analgesia are well recognized—such as the ability to administer analgesic drugs intermittently or continuously over several days, and to do so in lower doses due to the proximity of the catheter tip to the target region—it is not without risks. These include potential catheter dislodgement, local inflammation, and contamination of the insertion site [21,22]. The short execution time, easier postoperative management, and lower cost make the single-injection ("one-shot") technique a valid alternative for cats undergoing major surgeries.

All the cats received 0.2 mL/kg of 0.5% ropivacaine combined with 0.1 mg/kg of morphine. With this volume, anesthesia remained stable in the majority of cases evaluated, and no notable side effects were observed. The efficacy of the volume was demonstrated even in very small cats (case 2 weighted 1 kg). The current study referenced previous research on lumbosacral and thoracic epidural anesthesia in dogs, particularly focusing on the administration of ropivacaine and levobupivacaine combined with morphine. Epidural injection volumes commonly range from 0.1 to 0.22 mL/kg [23]. In a study by Lardone et al. [6], 0.2 mL/kg of 0.5% ropivacaine with morphine resulted in TEA without significant adverse effects, although prolonged sedation was reported in one case. This aligns with findings from Tonge et al. [7], where TEA with 0.5% levobupivacaine and morphine provided effective intraoperative analgesia without the need for additional systemic pain relief. In refrigerated cats, 0.2 mL/kg of diluted contrast agent resulted in an adequate longitudinal spread for effectively covering the thoracic and cranial lumbar spine [11].

The retrospective nature of this study inherently presented certain limitations, including incomplete data, variability in operators, and lack of standardization in perioperative management. The relatively low frequency of major thoracic surgeries in cats further contributes to the scarcity of studies on thoracic epidural anesthesia [23].

4. Conclusions

This case series demonstrated that TEA was feasible in cats. At the administered volume (0.2 mL/kg), TEA provided not only effective intraoperative analgesia but also good hemodynamic stability in cats undergoing major surgery.

Future prospective clinical studies involving larger populations and greater standardization of procedures are needed to further evaluate the intraoperative efficacy, potential side effects, and to characterize the extent and duration of postoperative analgesia provided by TEA in cats.

Author Contributions: E.L.: data acquisition, analysis, and interpretation; writing-review and editing, A.L.: acquisition of data and writing, G.M.: acquisition of data and writing, P.F.: conceptualization, review, and supervision. All authors have read and agreed to the published version of the manuscript.

Funding: This research received no external funding.

Institutional Review Board Statement: Ethical review and approval were waived for this study, due to its retrospective and clinical nature.

Informed Consent Statement: Written informed consent was obtained and signed from the owners of all the animals.

Data Availability Statement: The data presented in this study are available on request from the corresponding author due to privacy reasons.

Conflicts of Interest: The authors declare no conflicts of interest.

References

1. McLeod, G.; Cumming, C. Thoracic Epidural Anaesthesia and Analgesia. *Contin. Educ. Anaesth. Crit. Care Pain* **2004**, *4*, 16–19. [CrossRef]
2. Li, Y.; Dong, H.; Tan, S.; Qian, Y.; Jin, W. Effects of Thoracic Epidural Anesthesia/Analgesia on the Stress Response, Pain Relief, Hospital Stay, and Treatment Costs of Patients with Esophageal Carcinoma Undergoing Thoracic Surgery: A Single-Center, Randomized Controlled Trial. *Medicine* **2019**, *98*, e14362. [CrossRef]
3. Freise, H.; Van Aken, H.K. Risks and Benefits of Thoracic Epidural Anaesthesia. *Br. J. Anaesth.* **2011**, *107*, 859–868. [CrossRef]
4. Kozian, A.; Schilling, T.; Hachenberg, T. Non-Analgesic Effects of Thoracic Epidural Anaesthesia. *Curr. Opin. Anaesthesiol.* **2005**, *18*, 29–34. [CrossRef]
5. Son, W.G.; Jang, M.; Jo, S.M.; Yoon, J.; Lee, I. The Volume Effect of Lidocaine on Thoracic Epidural Anesthesia in Conscious Beagle Dogs. *Vet. Anaesth. Analg.* **2015**, *42*, 414–424. [CrossRef]
6. Lardone, E.; Sarotti, D.; Giacobino, D.; Ferraris, E.; Franci, P. Thoracic Epidural Anaesthesia vs Intrathecal Morphine in Dogs Undergoing Major Thoracic and Abdominal Surgery: Clinical Study. *BMC Vet. Res.* **2022**, *18*, 200. [CrossRef]
7. Tonge, M.; Robson, K.; Alderson, B. Single Thoracic Epidural Injection for Intra- and Post-Lateral Thoracotomy Analgesia in a Dog. *Vet. Rec. Case Rep.* **2021**, *9*, e183. [CrossRef]
8. Pageat, P. Stress During Hospitalization in Dogs and Cats: Consequences and Prevention. *J. Vet. Emerg. Crit. Care* **2021**, *31*, 629–637. [CrossRef]
9. Mesbah, A.; Yeung, J.; Gao, F. Pain After Thoracotomy. *BJA Educ.* **2016**, *16*, 1–7. [CrossRef]
10. Franci, P.; Leece, E.A.; Corletto, F. Thoracic Epidural Catheter Placement Using a Paramedian Approach with Cephalad Angulation in Three Dogs. *Vet. Surg.* **2012**, *41*, 884–889. [CrossRef]
11. Franci, P.; Santoro, F. Evaluation of the thoracic epidural spread after caudal thoracic contrast medium injections in cat cadavers. *Front. Vet. Sci.* **2025**, *12*, 1632175. [CrossRef]
12. Reid, J.; Scott, E.M.; Calvo, G.; Nolan, A.M. Definitive Glasgow Acute Pain Scale for Cats: Validation and Intervention Level. *Vet. Rec.* **2017**, *180*, 449. [CrossRef]
13. Adami, C.; Casoni, D. Addition of magnesium sulphate to ropivacaine for spinal analgesia in dogs undergoing tibial plateau levelling osteotomy. *Vet. J.* **2016**, *209*, 163–168. [CrossRef] [PubMed]
14. Palomba, N.; Marolf, V.; Spadavecchia, C.; Müller, N.; Sandersen, C.; Rohrbach, H. Peripheral Nerve Block versus Systemic Analgesia in Dogs Undergoing Tibial Plateau Levelling Osteotomy: Analgesic Efficacy and Pharmacoeconomics Comparison. *Vet. Anaesth. Analg.* **2020**, *47*, 119–128. [CrossRef]
15. Marolf, V.; Spadavecchia, C.; Müller, N.; Sandersen, C.; Rohrbach, H. Opioid Requirements after Locoregional Anaesthesia in Dogs Undergoing Tibial Plateau Levelling Osteotomy: A Pilot Study. *Vet. Anaesth. Analg.* **2021**, *48*, 398–406. [CrossRef]
16. Sarotti, D.; Lardone, E.; Piras, L.; Mancusi, D.; Franci, P. A Comparison of 0.375% Ropivacaine Psoas Compartment Block and 2% Prilocaine Spinal Anaesthesia in Dogs Undergoing Tibial Plateau Levelling Osteotomy. *BMC Vet. Res.* **2022**, *18*, 172. [CrossRef] [PubMed]
17. Sertcakacilar, G.; Tire, Y.; Kelava, M.; Nair, H.K.; Lawin-O'Brien, R.O.; Turan, A.; Ruetzler, K. Regional Anesthesia for Thoracic Surgery: A Narrative Review of Indications and Clinical Considerations. *J. Thorac. Dis.* **2022**, *14*, 5012–5028. [CrossRef]
18. Lee, J.Y.; Noh, K.M.; Lee, S.W.; Cho, E.Y.; Hong, J.H. Analysis of Factors Affecting Thoracic Epidural Anesthesia Performance Time. *Keimyung Med. J.* **2020**, *39*, 33–37. [CrossRef]
19. Waurick, R.; Van Aken, H. Update in Thoracic Epidural Anaesthesia. *Best Pract. Res. Clin. Anaesthesiol.* **2005**, *19*, 201–213. [CrossRef] [PubMed]
20. Valverde, A. Epidural analgesia and anesthesia in dogs and cats. *Vet. Clin. North Am. Small Anim. Pract.* **2008**, *38*, 1205–1230. [CrossRef] [PubMed]
21. DeRossi, R.; Hermeto, L.C.; Jardim, P.H.A.; de Andrade Bicudo, N.; de Assis, K.T. Postoperative Pain Control in Cats: Clinical Trials with Pre-Emptive Lidocaine Epidural Co-Administered with Morphine or Methadone. *J. Feline Med. Surg.* **2016**, *18*, 882–888. [CrossRef] [PubMed]
22. Hansen, B.D. Epidural Catheter Analgesia in Dogs and Cats: Technique and Review of 182 Cases (1991–1999). *J. Vet. Emerg. Crit. Care* **2001**, *11*, 95–103. [CrossRef]
23. de Paula Serrano, B.; Fukushima, A.R.; Muñoz, J.W.P.; Nascimento, F.L.D.; Conceição, V.D.S.F.; Almeida, G.R.; Almeida, T.R.; de Mello Morais, C.L.; Futema, F. Advancements in Epidural Anesthesia for Small Animals: A Comprehensive Literature Review. *J. Dairy Vet. Anim. Res.* **2024**, *13*, 1–10. [CrossRef]

veterinary sciences

MDPI

Article

The Dual Character of Animal-Centred Care: Relational Approaches in Veterinary and Animal Sanctuary Work

Anna K. E. Schneider [1] and Marc J. Bubeck [2,*]

1 FAU Kompetenzzentrum für interdisziplinäre Wissenschaftsreflexion, Friedrich-Alexander University Erlangen-Nürnberg, 91054 Erlangen, Germany; anna.schneider@fau.de
2 Faculty for Health Sciences Brandenburg, University of Potsdam, 14469 Potsdam, Germany
* Correspondence: marc.bubeck@uni-potsdam.de

Simple Summary: Caring for animals is central to veterinary and animal sanctuary work. However, the meaning of animal-centred care is often the subject of debate. This article explores the challenges faced by professionals when balancing two approaches: treating animals as individuals and considering their species-specific needs. These tensions are further shaped by the differing expectations of various groups—including animal owners, caretakers, veterinarians, institutions, and the public—who may hold conflicting views on what constitutes appropriate care. This article focuses on human–animal interactions in sanctuaries and the difficult decisions regarding killing an animal in veterinary medicine, such as euthanasia. The two care approaches and the expectations of actor groups can conflict—particularly when professionals must make decisions in real-life situations that are constrained by limited resources and time. By highlighting these tensions, this study provides insight into the nuances of animal-centered care and the skills needed by professionals to navigate the ever-increasing complexities of the field.

Abstract: Caring for the lives and welfare of animals is central to veterinary and animal sanctuary work, yet the meaning remains a subject of complex debates. Different stakeholders negotiate what constitutes appropriate care, leading to conflicting demands and expectations from internal and external sources. This article is based on two qualitative studies: Study I explores the multifaceted aspects of death work in farm animal medicine, emphasising the practical, emotional and ethical challenges involved. Study II examines human–animal interaction in sanctuaries, which reveal tensions between instrumental and relational care in animal-centred work. Relational care represents a subjectifying approach with individual attention to animals, while instrumental care is a more objectifying perspective based on species representation. These demands can often be contradictory, complicating day-to-day decision making under pressure. To analyse these complexities, this study employs Clarke's situational analysis (social worlds/arenas mapping), providing a means of comparing care work across different fields. This approach highlights how actor constellations, institutional settings, and structural constraints influence the negotiation of care. Addressing these issues provides a more nuanced understanding of the professional challenges of animal-centred care and the necessary skills to navigate its inherent contradictions.

Keywords: animal rights; animal welfare; care ethics; euthanasia; qualitative methods; working with animals

1. Introduction

Caring for the lives and well-being of animals is central to the professions of veterinary and animal-sanctuary work. In both fields, the meaning of animal-centred care is the subject of complex debates. For example, what are an appropriate space and structure for tigers in captivity or how to decide whether to treat or kill an animal in veterinary care. Different actors negotiate what constitutes appropriate care for animals in different ways. As a result, people who work professionally with animals face conflicting demands and expectations. These can come from within, from organisations and professional groups themselves, or from outside, from politics, the media or social movements. In addition to conflicting expectations, people working with animals have to make welfare (The term 'animal welfare' is understood in this text as the endeavour of ensuring the well-being of animals; it considers the facts that the term is viewed as scientifically controversial, interpreted in different ways, and critically discussed) decisions every day, often under less-than-ideal conditions, such as time pressure or limited financial or human resources [1,2]. These decisions reflect different forms of care: instrumental care, which focuses on ensuring species-appropriate standards and organisational efficiency, and relational care, which is shaped by empathy, trust, and close interaction with individual animals. What is considered appropriate care is not only negotiated among actors but also shaped by broader cultural and historical contexts. The meaning of good care is always embedded in specific social constructions of human–animal relationships, which shift over time and across societies [3–5]. These dimensions should therefore be an integral part of how we understand professional animal work.

Care work with animals is closely linked to questions of animal ethics. The discussion centres on safeguarding the welfare of animals. This can be viewed both from the perspective of *animal rights* and from the perspective of *animal welfare*. Animal welfare refers to an individual's state in coping with its environment, encompassing physical health, emotional states, and adaptive capacities [6]. Welfare assessments are based on measurable conditions and not merely on whether living circumstances are 'natural'. This broader scientific understanding provides an important framework for evaluating animal-centred care practices. While animal rights emphasise the subjective rights of animals, animal welfare advocates for better, more species-appropriate animal husbandry. In terms of content, there are parallels here with the care work explained above. The American Veterinary Medical Association defines animal welfare as the humane responsibility for the well-being of animals, including housing, management, disease prevention, care, and (if necessary) euthanasia [7]. This definition reflects a primarily instrumental understanding of care, where structured provisions are made for animals' physical needs. In contrast, ethical approaches, such as that of Palmer and Sandøe [8], emphasise the promotion of individual autonomy and development, aligning with what can be described as a relational understanding of care. These perspectives are also reflected in Nussbaum's capability approach, which argues for a dignified life for nonhuman animals that goes beyond basic welfare and includes emotional and developmental well-being [9].

When considering care work for animals and the associated practices, the understanding of animal welfare must be addressed. Institutional and individual interpretations influence the ways in which animals are cared for and in which care work is considered necessary. Care ethics offer a theoretical approach to the context-sensitive analysis of professional human–animal relationships [10]. Care encompasses a broad spectrum of behavioural practices that can also have negative effects [11]. For both professional groups, these practices are crucial to their understanding of their roles and shape their interactions.

In addition to the individual actors involved in animal-centred care work, the institutional setting must also be considered closely. The influence of institutions begins with the categorisation of animals based on laws or internal guidelines. Animals categorised as companion animals generally benefit from stronger protections than those classified as livestock. This disparity prompts ethical concerns about how different species are valued and the consequences for care practices. For example, farm animals are often treated with less regard than pets, sparking discussions about the moral implications of these distinctions [12]. These classifications influence the rights and protection of animals in everyday life and determine which actions are considered permissible, particularly in care work. In most cases, veterinary clinics and animal welfare organisations set standards that regulate interactions with animals and define treatment options.

Furthermore, institutional norms and cultures characterise everyday working life and determine the availability of time and financial resources. For example, tight schedules and limited resources in veterinary clinics increase pressure on staff, while financial constraints in animal welfare organisations limit the capacity to care for animals. Hierarchies and role allocations also have an influence on work with animals.

In this article, we explore the tensions inherent in professions that work with and care for animals. In doing so, the multiple stakeholders and expectations that shape animal-centred care are taken as our starting point. The argument is based on two major empirical studies: Study I examines the multi-dimensional aspects of killing in farm animal medicine, highlighting the ethical, emotional and practical demands placed on practitioners [13]. Study II focuses on human–animal interactions in different contexts, particularly in sanctuaries, with a particular emphasis on the challenges and complex negotiations inherent in caring [14]. These empirical perspectives allow us to trace how instrumental and relational care are enacted, justified, and negotiated in two highly contrasting institutional settings. To analyse these tensions, we draw on Clarke's [15] situational analysis and the social worlds/arenas theory developed by Strauss [16]. This theoretical framework offers tools for analysing how groups interact within specific social domains. Social worlds are dynamic, permeable groupings of actors connected through shared activities and identities, continuously shaped by processes of interaction and negotiation. Social arenas are understood as focused spaces of negotiation in which different social worlds (conflictually) exchange views on topics to assert their interpretations and, in some cases, assert them against others. The theory is characterised by the multi-perspectivity of social situations. This approach allows us to conceptualise animal-centred care as a negotiated practice shaped by overlapping actor constellations, institutional logics, and material conditions.

Through the comparative analysis, we demonstrate how the various expectations of stakeholders, including caretakers, veterinarians, animal owners, and government bodies, give rise to a complex arena in which animal-centred care work is negotiated. Building on this, we analyse how these external demands intersect with the dual approach of care in practice: instrumental care, which is oriented towards species-level objectivity, and relational care, which emphasises individualised, empathic engagement. This article provides an in-depth understanding of professional animal-centred care work and the complex negotiations involved, demonstrating how diverse stakeholder expectations intersect with the dual approaches of care, thereby intensifying the ethical and practical dilemmas faced by those in animal-related professions.

2. Materials and Methods

The present article employs a comparative approach to explore the complexities of animal-centred care work across two institutional contexts: sanctuaries and veterinary medicine. Utilising a qualitative research methodology, the care dynamics, responsibilities and ethical dilemmas that practitioners encounter in these two settings are analysed. The comparison allows not only for the specific challenges within each context to be illuminated but also for a more extensive understanding of how stakeholder expectations, institutional structures, and resource constraints influence care practices and interactions.

To achieve this objective, a range of qualitative research methods was employed, including semi-structured interviews and ethnographic observations, with a view to gathering rich, nuanced data. These methodologies facilitated the unearthing of the intricacies inherent in care work, encompassing the frequently disregarded emotional burdens and moral considerations that practitioners are required to navigate. In the context of care work, practitioners are continually interpreting the actions, needs and circumstances of other actors. Given the plurality of actors involved in care work—each bringing their own situational definitions and knowledge frameworks—an ecological approach is adopted to analyse these *"ecologies of knowledge"* [17]. This perspective emphasises the relationships and interdependencies within animal-centred care work, exploring how different stakeholders contribute to and negotiate the meaning of care [15].

To illustrate these dynamics, we draw on two extensive empirical studies: Study I examines the multifaceted aspects of killing in veterinary medicine, focusing on how veterinarians navigate the boundary work of determining what constitutes a 'good' or appropriate death. This research spans various practices, motives, and fields within veterinary medicine, such as small and large animal care. Study II explores human–animal interactions in sanctuaries, emphasising the complex relational dynamics that arise in these settings. Collectively, these studies enhance our comprehension of the multifaceted actors involved in care work and the intricate interplay between instrumental and relational care. Following the presentation of these individual study designs, the comparative approach is introduced (for an overview, see the COREQ checklist [18] in the Supplementary Materials).

2.1. *Study Design I: Veterinary Medicine*

In the field of veterinary medicine, daily activities encompass the care and treatment of animals, in addition to their killing [19,20]. This empirical study examines the intertwining of death work and veterinary professionalism, with a particular focus on the co-constitution of professional identity and ethical boundary making within veterinary practices [13]. This study adopts a grounded theory (GT) approach [21], exploring these complexities inductively with a view to contributing to the relatively underexplored field of sociological research on veterinary work [22]. Given the relative paucity of comparative sociological research on veterinary practices, this study is exploratory in nature [23]. The challenges associated with smaller sample sizes in comparative designs are mitigated by grounded theory principles such as constant comparison and theoretical sampling, which prioritise depth and theoretical relevance over representativeness.

2.1.1. Data Collection

The data collection process was conducted in accordance with the principles of constructivist grounded theory [21]. To capture both minimal and maximal contrasts within the sample, 17 veterinarians with diverse specialisations and disciplines were interviewed. The participants represent a broad spectrum of practice areas (e.g., small animal medicine, large animal medicine, exotic species), years of experience, specialties, gender, ages, geographic regions, and socioeconomic backgrounds.

Semi-structured interviews were conducted between 2020 and 2022 and lasted between 45 min and 2 h [24,25]. The interviews explored topics such as daily routines, killing practices, ethical dilemmas, and the professional biographies of veterinarians (see the Supplementary Materials). To complement the interviews, relevant documents such as guidelines, manuals, and websites were analysed to capture the discursive and material dimensions of killing practices within veterinary medicine.

2.1.2. Data Analysis

The analysis combined Charmaz's [21] constructivist grounded theory methodology with Clarke's [15] situational analysis (SA) to ensure a nuanced and comprehensive interpretation of the data. MAXQDA 24 software was utilised for systematic coding, thereby facilitating the identification of emergent themes and patterns. In the first stage, inductive open coding was applied to the interview transcripts. These were read multiple times in order to identify meaningful units within words, phrases, and sentences. These units were then given conceptual labels (codes) that reflected the participants' professional experiences. Similar codes were then grouped into broader conceptual categories. The initial coding process was guided by an awareness of the specific field of veterinary work. In the second phase, focused coding was employed to summarise and refine the material by identifying the most significant and recurring categories. This enabled a more abstract understanding of the professional logic of care to be achieved. Finally, theoretical coding integrated the findings with concepts within interdisciplinary debates in the sociology of professions, human–animal studies, and veterinary ethics. The study employed SA mapping strategies, including situational, social worlds/arenas, and positional maps, to visually represent the diverse and complex field of veterinary death work [26].

The establishment of interdisciplinary interpretation groups was a key aspect of the study, with the objective being to enhance the rigour of the analysis. The focus of these groups was on the validation of coding, as well as mapping and the refinement of emerging theoretical insights. This collaborative process enhanced intersubjectivity and ensured a thorough interpretation of the data.

2.1.3. Research Reflexivity and Ethics

The dual positionality of the researcher, who is both a veterinarian and a sociologist, provided unique access and profound field insights. This unique vantage point enabled an in-depth comprehension of the multifaceted professional and ethical complexities inherent in killing practices. However, this positionality also gave rise to challenges related to potential biases. To address these challenges, the researcher engaged in continuous reflexivity, documenting reflections throughout the research process and aligning the study with the interpretive goals of sociological inquiry.

Ethical approval was obtained from the Research Ethics Committee of the Faculty of Social Sciences, LMU Munich (GZ 23 13). All participants provided written informed consent, ensuring ethical standards were upheld.

2.2. *Study Design II: Rescue Centre*

This study examines care work in animal sanctuaries, specifically focusing on a 'big cat' sanctuary, to explore the dynamics of human–animal interactions and the roles of various stakeholders in care practices. Sanctuaries provide a unique setting for understanding how caregiving is shaped by both institutional structures and the interpersonal relationships between humans and animals. The ethnographic field study, conducted between 2019 and 2020, included visits on public open days and during non-public hours to observe caretakers' routines. Interviews combined spontaneous conversations on open days with scheduled appointments outside public hours. The study included interviews with six randomly selected volunteers to explore their backgrounds and motivations. At the time, two permanent caretakers were employed, and a third, in the hiring process, was unexpectedly available for an interview. Informal conversations with a visiting veterinarian and a former intern further supplemented the data.

2.2.1. Ethnographic Study

The ethnographic interviews, integrated with observations, were semi-structured to encourage more natural dialogue [27]. As rapport developed between the researcher and participants, this flexible format maintained the authenticity of the interactions. This approach is in line with Mayring's [28] recommendation to adapt methods to the research subject rather than rigidly applying pre-defined structures. Volunteers and carers were actively engaged in tasks, and interviews often took place alongside their regular duties, sometimes interrupted by responsibilities. Informal conversations were beneficial, allowing interviews to emerge naturally from ongoing interactions. The interview guide remained flexible, with questions evolving in response to the flow of the conversation.

Participant observation complemented the interviews by providing deeper insight into animal-centred care practices. According to Atteslander [29], participant observation involves "*systematically capturing, recording, and interpreting perceptible behavior at the time of its occurrence*" (p. 73). A key feature of this method is the researcher's personal involvement in the participants' interactions. As Mayring [28] explains, researchers become part of the social system they observe. Participation varied from full involvement in work tasks to observing caregivers' routines [30]. The researcher's role remained transparent, and participants were aware that the observations complemented the interview data. Observations were conducted across four different contexts and with varying levels of active participation: (1) open days, (2) active participation in a work assignment, (3) accompanying caretakers in their daily routines, and (4) observing caretakers and the veterinarian during vaccinations. These observations were not standardised but were guided by a clear focus on human–animal interactions and care relationships [31].

2.2.2. Data Analysis

Data analysis is based on a comprehensive dataset of semi-structured interviews, ethnographic field notes, visual materials (including photographs), and reflective memos. This approach provides a rich understanding of participants' experiences and their context. Interviews were audio-recorded, transcribed, and thematically analysed using MAXQDA 12 software. Initial coding followed a grounded theory approach, revealing key themes that reflect the complexity of human–animal care work in the sanctuary. Ethnographic field notes added context, highlighting social interactions and care practices, and were analysed using the same thematic framework to identify patterns consistent with the interview data. Thick descriptions captured the emotional and relational dynamics of care work, while visual material was analysed alongside verbal data to deepen the understanding of participants' experiences. Reflective memos documented the researcher's evolving thoughts and biases, enriching the analysis by reflecting on the dynamics between participants and the challenges of care work.

Building on these findings, social worlds/arenas mapping was employed to visualise the relationships between different actors and arenas in the sanctuary setting. This method helped create a map representing the interconnected social worlds and the dynamics that shape animal-centred care work. The map synthesises both instrumental and relational aspects of care work, offering a clear, spatial representation of how different roles and structures interact. The synthesis of these strategies provides a nuanced understanding of the interplay between instrumental and relational care work in the sanctuary, highlighting the importance of context and interpersonal relationships in shaping care practices.

2.3. Comparative Approach Using Social Worlds/Arenas Mapping

Social worlds/arenas theory provides a robust framework for investigating the complexities of professional practices involving animals within institutional settings. By emphasising the interactions and negotiations between diverse actors, this approach allows for a detailed exploration of how multiple stakeholder perspectives, institutional norms, and material conditions shape professional care practices.

To operationalise this framework, Clarke's [15] situational analysis (SA) was employed as a methodological tool. SA integrates the principles of grounded theory with mapping strategies, thereby enabling the systematic visualisation and analysis of relations within complex situations. Specifically, social worlds/arenas maps were constructed based on empirical data from the two case studies. These maps focus on the primary arena of animal-centred care work, with the various social worlds involved surrounding it. The maps represent relationships and attributes such as boundary permeability (e.g., dashed lines for porous boundaries) and size, with the latter indicating empirical prominence.

The subsequent synthesis of the data was facilitated by Clarke's mapping strategies, resulting in the creation of separate maps for each case study to identify key actors, relationships, and institutional structures. These maps were then used as analytical tools to explore how stakeholder expectations and institutional norms shape care practices.

The comparative approach was designed to synthesise findings from the two case studies, focusing on shared challenges and institutional distinctions. The maps enabled the identification of overlapping social worlds and arenas, allowing for the abstraction of broader dynamics that define animal-centred care work. Through this process, we explored how the duality of care—instrumental and relational—is shaped and intensified by external and internal demands across these professional contexts.

The integration of qualitative methods with situational analysis (social worlds/arenas mapping) enabled the study to transcend mere descriptive analysis, thereby providing a conceptual framework for comprehending the tensions and ethical dilemmas that are intrinsic to animal-centred care work. This methodological integration provides a rigorous basis for addressing the central aim of the paper, which is to illuminate how institutional structures and stakeholder negotiations co-construct the dual responsibilities of care in animal-oriented professions.

3. Study I: The Dual Approaches of Care in Veterinary Work

In veterinary medicine, caring for animals involves not only healing but also killing, which has an ethically ambivalent role. Veterinarians often perceive killing as a form of care, particularly when it is used for the purpose of alleviating suffering or for ethically justified purposes [20,32,33]. Veterinary care operates within a dynamic interplay of societal human–animal relationships, institutional frameworks and professional interactions [34–36]. A complex interplay shapes how care is understood, legitimized, and provided. Today, veterinary medicine is a specialized profession with various practices, settings, and roles. Farm animal medicine differs greatly from companion animal medicine because farm animals are considered not only patients but also production units [37]. These social categorizations influence the goals of veterinary care, shifting the focus from the health and well-being of patients to functionality and economic prosperity. This ambivalence is reflected in—and influences—the negotiation of relational and instrumental approaches to care.

3.1. Contextualizing Care in Farm Animal Medicine: Legal and Social Ambivalences

The ambiguity of the animal is reflected in its legal status in Germany. According to the German Civil Code (§90a BGB), animals have a special legal status that is distinct from property but not quite that of a legal subject. This tension between object and subject status exemplifies the ambivalent social perception of animals. Further distinctions, such as between livestock and companion animals, illustrate the legal nuances in veterinary practice. Horses, for example, occupy a unique position within the law and veterinary care [38]. Depending on their classification as food-producing animals, there are restrictions on the use of medicines that have a significant impact on their medical treatment. This classification highlights how the legal and institutional framework influences care practices and the ethical challenges faced by veterinarians.

In addition to legal issues, veterinary care is further complicated by the expectations of various stakeholders. These include media portrayals, the demands of animal welfare organisations, and the advocacy of law reform movements. A vet working with farm animals spoke about the challenges of balancing the demands of different parties. She talked about the negative portrayal of farm animal medicine in the media, as well as legal restrictions and the focus on profitability in this field: *"We've got a bit of a bad reputation, to be honest. There was a really nasty article in DIE ZEIT that called us 'drug dealers in green coats', which is just not true. On the one hand, we've got major restrictions on the active ingredients we're allowed to use for farm animals, and on the other, we've got to meet strict diagnostic requirements. The more we do, the more difficult it is to balance being economical and legal."* (veterinarian interview). Each stakeholder brings different and sometimes conflicting priorities to the table, adding to the complex landscape in which veterinarians operate.

Veterinary care is further complicated by its triadic structure, involving the veterinarian, the animal, and the client. Ethical tensions arise when the interests of these parties diverge, particularly in matters of prioritisation [39,40]. The role of the veterinarian is often debated: Is their function akin to that of a mechanic, whose duty is to follow the client's instructions, or closer to that of a paediatrician, who advocates for the best interests of the dependent patient?

Law [41] extends this framework by suggesting two additional *"objects of care"* that influence veterinarians: the wider societal responsibilities ('the bigger picture') and the veterinarians themselves. These levels illustrate how the demands on veterinarians vary from case to case, shaped by the complex web of actors, institutions, and societal expectations.

All of the interviewed vets emphasised that their work is always about the animal: *"I like (...) working with animals, I think they're just very, very pleasant creatures"* (veterinarian interview). Despite this, safe food production was another central motive, as exemplified by the statement of one vet: *"I like being around people and animals. And working in food production makes sense to me, it's something relevant and important to us as a society"* (veterinarian interview). Veterinary work is not only imbued with the social values of caring for animals but also of producing safe food: "Even this work in agriculture, I just see as something that is relevant, something that is important for us as a society" (veterinarian interview). This balancing of priorities often places veterinarians in ethically complex situations where food safety and economic considerations influence care decisions, particularly in relation to euthanasia.

3.2. Balancing Instrumental and Relational Care

In the context of farm animal medicine, instrumental care refers to interventions aimed at preserving animal health and functionality within a production system, which often prioritizes productivity, biosecurity, and herd management. In contrast, relational care focuses on the well-being of individual animals and the emotional and ethical connections among veterinarians, animals, and owners.

A farm veterinarian provided a detailed account of the treatment of a cow suffering from milk fever, a condition that required immediate action to save the cow's life and preserve her productivity. The veterinarian's intervention was guided by expert knowledge of the disease's pathophysiology and the practical goal of restoring the animal's productivity. While undoubtedly lifesaving, this care primarily reflects an instrumental approach in which the cow's value is closely tied to her function within the production cycle. Relational aspects, such as the veterinarian's emotional connection to the animal or the animal's individual experience, remain implicit or subordinated. This points to structural constraints that discourage deeper individual engagement. This is illustrated by the description of a female veterinarian, in which the physiological state of the animal is in the foreground and the act of killing is portrayed as a relief: *"For me, it wasn't too bad because the animal was only released. I saw the whole story and it was hopeless, and the animal had suffered a lot. It was a trapped cow, and for me, it was really about simply releasing the animal from its suffering"* (veterinarian interview). However, the interviews also include examples in which animals are euthanised because appropriate care is not possible on the farm. In such so-called *"animal welfare cases"* (veterinarian interview), veterinarians must make decisions that reveal the limitations of both instrumental and relational care. These cases are perceived as particularly distressing. These cases take an emotional toll on veterinarians, reflecting their ethical ambivalence and tension. Deelen et al. [42] highlighted the incongruity between ethical principles, such as the 'best interest' of the animal, and the pragmatic realities that veterinarians encounter in their daily practice.

Another veterinarian described a situation where she treated a farmer's *"favourite cow"* (veterinarian interview). Despite her best efforts, the cow's health deteriorated due to complications from a claw resection, and the cow was ultimately sent to the slaughterhouse, a decision the veterinarian later regretted. She expressed her remorse, stating, *"In hindsight, I wish we had euthanised her earlier. The journey to the slaughterhouse could have been spared; it was not necessary"* (veterinarian interview). This case exemplifies the conflicting demands of care: the veterinarian's emotional engagement with the individual animal and the institutional pressures that prioritise the cow's classification as a productive resource. The veterinarian's reflections underscore the tension between relational care, grounded in empathy for the animal, and the structural realities of farm animal medicine, which often limit such considerations. The *"favourite cow"* case is an example of a rare moment when relational care takes centre stage. The veterinarian treats the animal not only as a patient or production unit but also as an individual with whom the farmer and the vet share a history. The subsequent regret about not euthanizing the animal earlier suggests dissonance between the veterinarian's emotional commitment and the institutional requirement to preserve economic value for as long as possible. These moments of moral distress reveal how relational care persists in the background and is often in tension with dominant expectations.

These cases demonstrate the intertwining of instrumental and relational care in farm animal medicine. Although care is typically provided within a framework that emphasizes productivity and efficiency, veterinarians also engage in emotionally charged, ethically complex practices that demonstrate their commitment to the well-being of individual animals. As one veterinarian insightfully observed, *"We're in a bind—our clients are the animal owners, but we're also responsible for the animals' welfare"* (veterinarian interview). The persistence of relational care, even under institutional constraints, reveals its enduring role in farm veterinary medicine. Recognizing this dual approach of care can improve our understanding of the emotional labour and moral tensions that shape everyday veterinary practice.

4. Study II: The Complex World of Care Work in Rescue Centres

While veterinary medicine highlights tensions between relational and instrumental care in a clinical setting, animal sanctuaries offer a different institutional and emotional landscape. Here, the data reveal complex dynamics between these two forms of care, shaped by nuanced human–animal interactions, as well as structural challenges. Caregivers operate within a framework shaped by institutional constraints, resource limitations, and the individual needs of the animals. These challenges reveal how the practical and emotional aspects of animal-centred care intersect, affecting both the carers and the animals.

4.1. Care Work in the Sanctuary: Context and Challenges

"The most important task as an animal carer? That my animals are well. Quite simply, that my animals are doing well" (caregiver interview).

The selected wildcat sanctuary operates with a mix of full-time staff, volunteers, and interns who care for animals from a variety of backgrounds, including zoos, circuses, and private ownership. Although the organisation sees itself as a temporary sanctuary, ideally with the goal of finding better permanent homes, many of the hard-to-rehome wild cats end up staying there permanently. New staff at the sanctuary, often trained in hierarchical zoo environments, undergo a transition to its more personalised yet administratively complex working structure, where unclear leadership, differences in expertise among team members, and chronic understaffing pose significant challenges. As one of the few full-time staff members summarised it, *"I feel more responsible here than at the zoo"* (caregiver interview). However, this increased responsibility comes with considerable challenges, as professional staff with specialist knowledge must work with volunteers who may lack comparable knowledge yet are actively involved in decision-making processes. This interplay reveals that relational care, grounded in responsibility and emotional commitment, may be disrupted by institutional or knowledge-based hierarchies—an often-overlooked aspect of animal-centred work. This dynamic often causes conflict or inefficiency due to mismatched expertise and authority within the sanctuary team. The sanctuary's outdated enclosures and financial constraints compound the challenges. While efforts to improve living conditions continue, these constraints affect both the welfare of the animals and the daily routines of their caretakers.

4.2. Settling In: Individualised Care and Challenges

The individualised care of animals at the sanctuary begins with their admission and acclimatisation. An example of this process is illustrated by an employee's account of a serval: *"With him [pointing to Serval] it's just a difficult story, because he came from this flat, that is, he was used to people, this woman also told us that he sometimes slept in bed with us [sic], [...] but when he was here, he didn't show himself to be accessible to us at all, so really zero"* (caregiver interview). In cases like this, a longer familiarisation process ensues, wherein staff members work to build trust through food and play. However, disruptive events—such as necessary veterinary check-ups or vaccinations—can hinder this progress, creating a delicate balance between granting the animal freedom and familiarising it with new living conditions. As one staff member notes, *"Well, when an animal is new here, you actually try to leave it as undisturbed as possible at the beginning, so that it can acclimatise in peace, i.e., to build up as little pressure and stress as possible, but on the other hand I at least do it in such a way that I show myself to the animal a lot, even during normal procedures, so that the animal also gets to know it, so don't wrap it in cotton wool, it should get to know the noises and everything that goes on here every day"* (caregiver interview).

This acclimatisation process is further complicated by the inherent dangers of the animals, which limits direct access for handling. For example, moving animals from outdoor to indoor enclosures is a necessary task due to legal housing requirements, cleaning schedules, or significant temperature drops. Additionally, some animals respond differently to individual carers or volunteers. Employee R describes a female cougar as a *"cat for women"* and notes that she is *"female-oriented"* (caregiver interview). This observation prompts a male colleague to invest extra effort in building rapport with her: *"I realised that the puma is afraid of me because she doesn't know me at all. She's mistrustful of men anyway, but she gets on really well with [staff member R] and not with me. So, I spend more time with the puma now, bribing her with a little milk [...] to show her that I'm not so bad"* (caregiver interview). Such relational care fosters trust, which is essential for the daily interactions between caregivers and the animals.

4.3. Instrumental Care: Practical Responsibilities

Instrumental care encompasses the practical, routine tasks required to maintain animal welfare, such as enclosure maintenance and health interventions. These practices are primarily guided by species-specific standards and legal requirements, often carried out under considerable time and resource constraints. Daily cleaning routines are complemented by periodic deep cleaning and the replacement of bedding, requiring considerable effort and volunteer involvement. For instance, staff must move logs, branches, and other materials to ensure the animals' environments remain enriching. While some changes provide enrichment, others can be challenging for individual animals, particularly those with specific needs, such as the caracal that is blind in one eye. To accommodate its spatial awareness difficulties, the staff has opted to maintain a stable habitat to allow the animal to navigate comfortably.

The sanctuary adheres to the principles of behavioural enrichment, incorporating various structures and toys to enhance the animals' daily lives. For example, after Christmas, staff collect discarded Christmas trees, whose unique scents and textures provide stimulating experiences. One caretaker reflects, *"So we make an effort, I'd like to do a bit more, but there's often not enough time to really sit down and prepare and make something more elaborate, which is why it's usually something that you just put in or hang up, which doesn't require much effort"* (caregiver interview). Due to limited resources, care that extends beyond basic provisioning often relies on external support, such as workshops with visitors and volunteers to create enrichment elements from materials like fire hoses or car tires.

Health care is another crucial aspect of instrumental care. The sanctuary collaborates with a local veterinarian who has experience with exotic animals. Routine veterinary visits, including vaccinations, require meticulous planning and preparation. For large cats like tigers and pumas, vaccinations can occur from a distance using a blowgun, while smaller animals necessitate direct injections. Staff utilise specially designed boxes to facilitate safe and close contact during these procedures, although such preparations require time and are often constrained by daily schedules and safety regulations.

Despite the necessity of veterinary interventions, these visits are often negative experiences for both animals and staff, compounded by the inability to communicate the underlying intentions of care: *"They don't realise that we're only doing them good, for them the negative experience, the prick, the stress, that just stays there"* (caregiver interview). This communicative barrier highlights a fundamental tension within relational care: while caregivers may act out of genuine concern, their inability to explain or justify these actions to the animals undermines the trust they seek to build. The resulting mismatch between intention and reception is not only emotionally taxing but also ethically charged, as it confronts the limits of empathy and understanding across species boundaries. This dimension becomes particularly relevant when comparing professional care in sanctuaries with veterinary contexts, where emotional bonds and practical interventions are continuously re-negotiated. Efficiency and pragmatic implementation take precedence, often forcing caregivers into ethically and emotionally challenging interactions; for example, caregivers may use sliding doors as lures to encourage animals closer to the blowgun. Staff often feel a palpable sense of resentment toward the stressful impact of these procedures on the already tense animals, and each successful vaccination is met with relief—not only for procedural success but in order to preserve a fragile relational dynamic.

4.4. Relational Care: Building Trust

In contrast, relational care emphasises the establishment of trust and understanding between the caregivers and the animals. Daily interactions are crucial for rebuilding and maintaining this trust, even amid intrusive medical measures. *"I had picked up a pretzel with butter from the bakery earlier and was eating it when I broke off a piece to give to her. Surprisingly, she was completely relaxed, even though we had just caught her half an hour earlier using significant restraint, and I had to hold her tightly to prevent her from biting. Despite that, she sat calmly in front of me looking out and taking the pretzel from me. I found it fascinating. People often say that we don't want to harm them, and perhaps she sensed that. Yet, the very act of capturing her with force could have made her want to withdraw completely. I found that really impressive"* (caregiver interview). This moment encapsulates a fundamental tension within caregiving practices: the caregiver's emotional investment and ethical intent may not always coincide with species-specific welfare protocols. The act of offering a buttered pretzel—perhaps motivated by a desire to comfort or reestablish connection—underscores the practical and ethical complexities of navigating affective responsiveness within the constraints of institutional care. Such seemingly minor gestures illuminate broader ambiguities inherent in sanctuary-based caregiving, where anthropomorphic inclinations, shifting care standards, and affective bonds frequently intersect. Rather than interpreting these actions solely as errors or lapses in judgment, we propose that they be understood as illustrative of the ongoing negotiation that defines care work—an arena marked by the continual recalibration of ethical ideals, material constraints, and the lived dynamics of interspecies relations. Caregivers recognise the downsides of their work, grappling with the potential consequences for their relationships with the animals, particularly concerning health-related interventions that may involve difficult decisions, such as euthanasia.

In the sanctuary, interactions between humans and animals are shaped by a physical division of their living spaces, highlighting themes of proximity and distance. Unlike pets that are integrated into human households, wild animals reside within human structures that are tailored to meet their needs. Consequently, these encounters occur in a liminal space that considers the animals' natural requirements alongside human factors, including architectural and financial constraints and practical caregiving issues. Following the social worlds/arenas theory, it becomes clear how many different social worlds interact in the realisation of such measures. Within the animal sanctuary, the carers' groups coordinate their activities with volunteers and association boards. At the same time, the worlds of visitors, institutions, and regulating authorities are also affected. The desire to improve husbandry conditions is complex and requires the purchase of land, which affects financing and utilisation requirements. Food donations also fluctuate in quantity and quality, influenced by various external actors, such as slaughterhouses and hunters.

In summary, the analysis of care work in the sanctuary reveals a nuanced interplay between instrumental and relational care. While instrumental care encompasses the practical aspects of animal welfare, relational care highlights the importance of building trust and understanding between carers and animals. Sanctuary staff strive to navigate these dimensions while meeting legal and public expectations, ultimately aiming to provide a safe and nurturing environment for the animals in their care. Sanctuary staff treat each animal as an individual from the moment they arrive, taking into account their specific needs and personality traits. Their training enables them to professionally assess and interpret animal behaviour, fostering an intimate understanding of each creature. However, this intimacy is balanced by the need for professional distance, especially during veterinary visits or rehoming. Caregivers must sometimes prioritise the wider interests of animal welfare over personal emotional relationships. Animal welfare legislation imposes further restrictions on care practices, requiring a balance between legal requirements and the needs of the animals. Public pressure for transparency and efficiency often leads to conflict and funding challenges. Therefore, effective sanctuary care requires recognition of these complexities and the development of strategies that balance financial support, legal frameworks, public perception, and internal communication.

5. Negotiating Dual Approaches of Animal-Centred Care Work

This comparative study explores the inherent tensions within professions dedicated to animal care. Based on two empirical cases from veterinary medicine and animal sanctuaries in Germany, we analysed how animal-centred care is shaped by negotiations between stakeholder expectations, institutional frameworks, and professional judgement. These care practices occur within a broader context in which various logics of action—economic, legal, relational, and ethical—intersect.

Our findings confirm two aspects of care: instrumental care, which is focused on standardised and efficient procedures that are appropriate for the species in question, and relational care, which is characterised by empathy, individual attention, and trust. Rather than existing as distinct approaches, these forms of care occur together in everyday practice. Professionals and caregivers must constantly adapt to situational demands and ethical complexities by navigating between them.

To visualise this complexity, we created a social worlds/arenas map (see Figure 1), which illustrates the variety of groups involved in animal-centred care work. Rather than aiming for completeness, the map emphasizes key tensions to facilitate comparative analysis of animal-centred care work. The central blue area represents the dynamic space in which care is provided. The surrounding, intersecting social worlds demonstrate how diverse institutions, professions, and movements influence the meanings and boundaries of care. The open worlds on the margins reflect the openness and fluidity of social processes, where boundaries are permeable and roles are continually evolving.

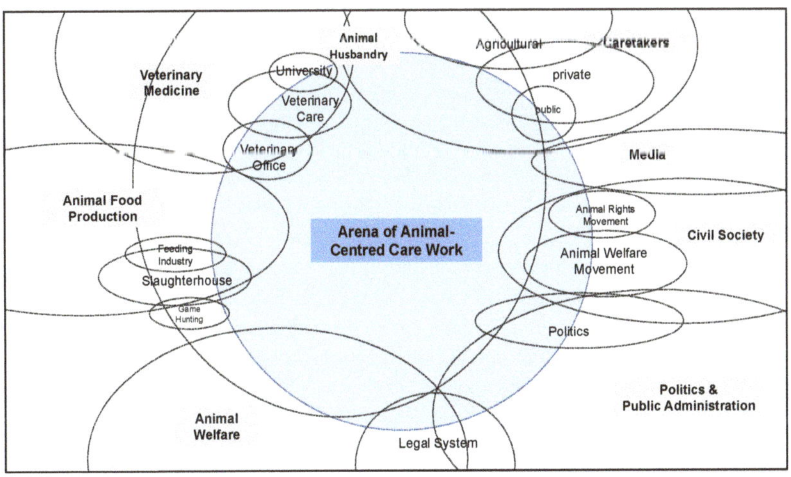

Figure 1. Arena of animal-centred care work.

Veterinary medicine is a prominent feature on the left, encompassing practices, clinics, and offices that are responsible for medical treatment, public health, and legal enforcement. This sector is closely intertwined with the fields of animal food production, including the feeding industry, slaughterhouses, and game hunting, where economic and legal demands dominate. Below this cluster, animal welfare emerges as a scientific and legal domain, bridging research, ethical discourse, and regulatory policy.

In contrast, the right side of the map is shaped by civil society actors, particularly animal welfare and animal rights movements. These groups exert influence on public discourse, shape legislation, and often challenge institutional practices. This sphere intersects with the media, which reflects and shapes societal attitudes towards animals. Also depicted are caretakers, whose roles span private homes, public institutions, and sanctuaries, each with different care principles and constraints. Finally, political and public administration fields connect and mediate between legal, scientific, and public demands.

This visualisation highlights a key point: care work is not carried out in isolation but is embedded in a web of competing expectations and value systems. The complexity of this arena necessitates a flexible and responsive approach to care that incorporates structural and relational considerations. For instance, veterinary professionals must comply with regulations and manage efficiency while responding empathetically to animal suffering. Similarly, sanctuary staff must balance intimate, trust-based relationships with practical constraints and safety rules.

Both study settings reveal an ongoing interplay between instrumental and relational care, which is influenced by institutional frameworks, societal expectations, and situational demands. Care is not purely objective or entirely empathetic but, rather, a dynamic practice that is negotiated within a given context. In farm animal medicine, instrumental care often dominates due to species-level objectives such as productivity, food safety, and cost efficiency [37]. However, this does not exclude relational elements: veterinarians report emotional moments such as regret over euthanasia decisions or recognition of animals as sentient beings [13,41]. These moments demonstrate that relational care can emerge even within structured, efficiency-driven systems. In rescue centres, relational care is more explicit [43]: caregivers build trust and form individual bonds with the animals in their care. Nevertheless, instrumental concerns remain central. Practical demands such as limited resources, safety protocols, and legal compliance necessitate a structured approach to care [44]. Even trust-building processes, such as acclimatisation, are influenced by institutional constraints. This interplay between institutional mandates and ethical imperatives often generates complex dilemmas, particularly in cases where an animal's well-being conflicts with prescribed protocols [45]. Together, these cases demonstrate that animal-centred care involves an ongoing negotiation between ethical sensitivity and pragmatic demands, with professionals continually adapting their approach.

While this study focuses on institutional and professional negotiations surrounding animal-centred care, it is important to acknowledge a further dimension: the role of animals as more than passive recipients of care. Although not the primary focus here, emerging research in animal studies and multispecies ethnography emphasises the significance of animal agency in shaping care relations. Moments of trust, resistance, or behavioural expression on the part of animals are not merely background conditions but can actively influence how care is enacted and experienced. Recognising this dimension further underscores the complexity of care as a negotiated and relational practice [11,41,46].

Therefore, care is always situated. It reflects the professional, cultural, and institutional environment in which it occurs. In the German context, for instance, this encompasses robust legal frameworks for animal welfare, relatively high levels of public awareness of animal rights, and substantial civil society involvement. While these findings may not be directly generalisable to other socio-legal contexts, they offer important insights into how care practices are embedded in broader societal frameworks. Ultimately, we argue that adequate, animal-centred care requires a continual negotiation between the instrumental and relational dimensions. Neither dimension is sufficient on its own. Only by acknowledging this duality can professionals and caregivers respond ethically and effectively to the complex and often contradictory demands of their work.

6. Conclusions

In light of these findings, the following conclusion summarises the key insights and implications for professional care work. Animal-centred care work in veterinary medicine and animal sanctuaries is characterised by two approaches: while instrumental care focuses on species-based needs, relational care emphasises the individual and relational aspects. Those working with animals are faced with the challenge of balancing ethical responsibilities with economic constraints and institutional conditions. Structural frameworks play a central role in shaping care practices by categorising animals, defining regulatory guidelines, and determining available resources. These structures reflect societal values and can either constrain or enable the quality of care. The valuation of the two care approaches is negotiated through a social process involving different actor groups. Addressing the dual demands of care requires not only policy reforms to ensure adequate resources, but also a cultural shift that recognises care as a core professional and ethical practice.

Supplementary Materials: The following supporting information can be downloaded at: https://www.mdpi.com/article/10.3390/vetsci12080696/s1, Overview of Both Study Designs; Interview Guide for Study I: Veterinary Medicine.

Author Contributions: Conceptualisation, A.K.E.S. and M.J.B.; methodology, A.K.E.S. and M.J.B.; software, A.K.E.S. and M.J.B.; validation, A.K.E.S. and M.J.B.; formal analysis, A.K.E.S. and M.J.B.; investigation, A.K.E.S. and M.J.B.; resources, A.K.E.S. and M.J.B.; data curation, A.K.E.S. and M.J.B.; writing—original draft preparation, A.K.E.S. and M.J.B.; writing—review and editing, A.K.E.S. and M.J.B.; visualization, A.K.E.S. and M.J.B.; supervision, A.K.E.S. and M.J.B.; project administration, A.K.E.S. and M.J.B.; funding acquisition, A.K.E.S. and M.J.B. All authors have read and agreed to the published version of the manuscript.

Funding: This research received no external funding. M.B.'s dissertation project was funded by the Studienstiftung des deutschen Volkes. A.S. was supported by the Dr. Alfred Vinzl Foundation and the Luise Prell Foundation, Erlangen, Germany.

Institutional Review Board Statement: The studies were conducted in accordance with the Declaration of Helsinki. The involvement of human participants followed the Code of Ethics of the German Sociological Association, strictly adhering to ethical principles, including participant consent, informing them of their rights, voluntary participation, data anonymization, and safeguarding their privacy. M.B.'s study was additionally approved by the Research Ethics Committee of the Faculty of Social Sciences, LMU MUNICH (GZ 23-13, 6 December 2023).

Informed Consent Statement: Informed consent was obtained from all subjects involved in the study.

Data Availability Statement: The anonymised data used to support the conclusions of this article can be provided by the authors.

Acknowledgments: We would like to thank the anonymous reviewer, as well as Clara v. Keusgen, Mina Mittertrainer, Alexander Smith, and Levi I. Ufferfilge, for their valuable critical engagement and thoughtful comments on earlier versions of this manuscript, as well as for their support with proofreading.

Conflicts of Interest: The authors declare no conflicts of interest.

References

1. Hamilton, L.; Taylor, N. *Animals at Work: Identity, Politics, and Culture in Work with Animals*; Brill: Boston, MA, USA, 2013; ISBN 978-90-04-23582-3.

2. Tallberg, L.; Jordan, J.J. Killing Them 'Softly' (!): Exploring Work Experiences in Care-Based Animal Dirty Work. *Work Employ. Soc.* **2021**, *36*, 843–859. [CrossRef]
3. Franklin, A. *Animals and Modern Cultures: A Sociology of Human–Animal Relations in Modernity*; SAGE Publications: London, UK, 1999; ISBN 978-1-4462-6454-6.
4. Arluke, A.; Sanders, C.R. *Regarding Animals*; Temple University Press: Philadelphia, PA, USA, 1996; ISBN 9781566394413.
5. DeMello, M. *Animals in Society. An Introduction to Human-Animal Studies*; Columbia University Press: New York, NY, USA, 2012; ISBN 9780231152945.
6. Broom, D.M. Welfare concepts. In *Encyclopedia of Animal Behavior*, 3rd ed.; Brown, C., Ed.; Elsevier: Amsterdam, The Netherlands, 2024. [CrossRef]
7. American Veterinary Medical Association. "Animal Welfare.". Available online: https://www.avma.org/resources-tools/animal-health-and-welfare/animal-welfare (accessed on 10 June 2025).
8. Palmer, C.; Sandøe, P. Critical Terms for Animal Studies. In *Critical Terms for Animal Studies*; Gruen, L., Ed.; University of Chicago Press: Chicago, IL, USA, 2018; pp. 424–438, ISBN 9780226355429.
9. Nussbaum, M.C. *Frontiers of Justice: Disability, Nationality, Species Membership*; Harvard University Press: Cambridge, MA, USA, 2006; ISBN 0674019172.
10. Collins, S. *The Core of Care Ethics*; Palgrave Macmillan: New York, NY, USA, 2015; ISBN 978-1-349-43645-3.
11. de la Bellacasa, M.P. *Matters of Care: Speculative Ethics in More than Human Worlds*; University of Minnesota Press: Minnesota, MN, USA, 2017; ISBN 978-1-4529-5347-2.
12. Fraser, D. Toward a Global Perspective on Farm Animal Welfare. *Appl. Anim. Behav. Sci.* **2008**, *113*, 330–339. [CrossRef]
13. Bubeck, M.J. Justifying Euthanasia: A Qualitative Study of Veterinarians' Ethical Boundary Work of "Good" Killing. *Animals* **2023**, *13*, 2515. [CrossRef] [PubMed]
14. Schneider, A.K.E. *Interaktion Zwischen Spezies: Fremdheit und Nähe*; Springer VS: Wiesbaden, Germany, 2023; ISBN 978-3-658-41894-6.
15. Clarke, A.E.; Friese, C.; Washburn, R.S. *Situational Analysis. Grounded Theory After the Interpretive Turn*, 2nd ed.; SAGE Publications: Thousand Oaks, CA, USA, 2018; ISBN 9781452260907.
16. Strauss, A. A social world perspective. In *Studies in Symbolic Interaction. An Annual Compilation of Research*; Denzin, N.K., Ed.; JAI Press: Greenwich, CT, USA, 1978; pp. 119–128, ISBN 978-0892320653.
17. Star, S.L. (Ed.) *Ecologies of Knowledge: Work and Politics in Science and Technology*; State University of New York Press: Albany, NY, USA, 1995; ISBN 978-0-7914-2566-4.
18. Tong, A.; Sainsbury, P.; Craig, J. Consolidated Criteria for Reporting Qualitative Research (COREQ): A 32-Item Checklist for Interviews and Focus Groups. *Int. J. Qual. Health Care* **2007**, *19*, 349–357. [CrossRef] [PubMed]
19. Dürr, S.; Fahrion, A.; Doherr, M.G.; Grimm, H.; Hartnack, S. Akzeptanz Des Tötens von Tieren: Umfrage bei Tierärzten und anderen Berufsgruppen. *Schweiz. Arch. für Tierheilkd.* **2011**, *153*, 215–222. [CrossRef] [PubMed]
20. Gardiner, A. Killing and Caring: A Historical Perspective in Veterinary Medicine. *Vet. Hist.* **2019**, *20*, 6–27.
21. Charmaz, K. *Constructing Grounded Theory*, 2nd ed.; SAGE: London, UK, 2014; ISBN 9780857029140.
22. Bonnaud, L.; Fortané, N. Being a Vet: The Veterinary Profession in Social Science Research. *Rev. Agric. Food Environ. Stud.* **2020**, *102*, 125–149. [CrossRef]
23. Anderson, A.; Hobson-West, P. Animal Research, Ethical Boundary-work, and the Geographies of Veterinary Expertise. *Trans. Inst. Br. Geogr.* **2022**, *48*, 491–505. [CrossRef] [PubMed]
24. Bogner, A.; Littig, B.; Menz, W. (Eds.) *Interviewing Experts*; Palgrave Macmillan: London, UK, 2009; ISBN 978-0-230-22019-5.
25. Bogner, A.; Littig, B.; Menz, W. (Eds.) *Das Experteninterview: Theorie, Methode, Anwendung*; Leske + Budrich: Opladen, Germany, 2002; ISBN 978-3-322-93270-9.
26. Bubeck, M.J.; Jansky, B. Relational und dimensional: Heuristische Dimensionen in der Situationsanalyse am Beispiel von Care-Praktiken und ihrer Materialität. *Forum Qual. Sozialforschung Forum Qual. Soc. Res.* **2023**, *24*. [CrossRef]
27. Davies, C.A. *Reflexive Ethnography: A Guide to Researching Selves and Others*; Routledge: London, UK, 2012; ISBN 978-0-203-06937-0.
28. Mayring, P. *Einführung in die Qualitative Sozialforschung Eine Anleitung zu Qualitativem Denken*, 5th ed.; Beltz Verlag: Weinheim, Germany, 2002; ISBN 9783407252524.
29. Atteslander, P.; Cromm, J. *Methoden der Empirischen Sozialforschung*, 9th ed.; De Gruyter: Berlin, Germany, 2000; ISBN 978-3-11-016835-8.
30. Lamnek, S. *Qualitative Sozialforschung: Methoden und Techniken*, 2nd ed.; Beltz: Weinheim, Germany, 1993; ISBN 3621271775.
31. Bortz, J.; Döring, N. *Forschungsmethoden und Evaluation für Sozialwissenschaftler*, 2nd ed.; Springer: Berlin, Germany, 1995; ISBN 3540593756.
32. Bubeck, M.J. Veterinarians as protectors of animals—Caring and killing in the narratives of vet students. In *Justice and Food Security in a Changing Climate*; Schübel, H., Wallimann-Helmer, I., Eds.; Wageningen Academic Publishers: Wageningen, The Netherlands, 2021; pp. 248–254, ISBN 9789086869152.
33. Morris, P. *Blue Juice: Euthanasia in Veterinary Medicine*; Temple University Press: Philadelphia, PA, USA, 2012; ISBN 9781439907054.

34. Swabe, J. *Animals, Disease and Human Society. Human-Animal Relations and the Rise of Veterinary Medicine*; Routledge: London, UK, 1999; ISBN 9780415181938.
35. Jones, S.D. *Valuing Animals: Veterinarians and Their Patients in Modern America*; Johns Hopkins University Press: Baltimore, MD, USA, 2002; ISBN 978-0-8018-7129-0.
36. Bubeck, M.J.; Springer, S. Sociologising Veterinary Ethics: An Exploration of Professional Identity. In *EurSafe2024 Proceedings. Back to the Future—Sustainable Innovations for Ethical Food Production and Consumption*; Giersberg, M., Meijboom, F., Bovenkerk, B., Eds.; Brill: Leiden, The Netherlands, 2024; pp. 324–329, ISBN 978-90-04-71550-9.
37. Springer, S.; Grimm, H. The Future of Veterinary Ethics: History, Diagnosis, and Prognosis of an Evolving Research Field. In *Ethics in Veterinary Practice: Balancing Conflicting Interests*; Kipperman, B., Rollin, B.E., Eds.; Wiley-Blackwell: Hoboken, NJ, USA, 2022; pp. 481–500, ISBN 9781119791195.
38. Vermilya, J.R. Contesting Horses: Borders and Shifting Social Meanings in Veterinary Medical Education. *Soc. Anim.* **2012**, *20*, 123–137. [CrossRef]
39. Kipperman, B. Veterinary Advocacies and Ethical Dilemmas. In *Ethics in Veterinary Practice: Balancing Conflicting Interests*; Kipperman, B., Rollin, B.E., Eds.; Wiley-Blackwell: Hoboken, NJ, USA, 2022; pp. 123–144, ISBN 9781119791195.
40. Tannenbaum, J. Veterinary Medical Ethics: A Focus of Conflicting Interests. *J. Soc. Issues* **1993**, *49*, 143–156. [CrossRef]
41. Law, J. Care and Killing: Tensions in Veterinary Practice. In *Care in Practice: On Tinkering in Clinics, Homes and Farm*; Mol, A., Moser, I., Pols, J., Eds.; Transcript: Bielefeld, Germany, 2010; pp. 57–69, ISBN 978-3-8376-1447-3.
42. Deelen, E.; Meijboom, F.L.B.; Tobias, T.J.; Koster, F.; Hesselink, J.-W.; Rodenburg, T.B. The Views of Farm Animal Veterinarians about Their Roles and Responsibilities Associated with On-Farm End-of-Life Situations. *Front. Anim. Sci.* **2022**, *3*, 949080. [CrossRef]
43. Veasey, J.S. In pursuit of peak animal welfare; the need to prioritize the meaningful over the measurable. *Zoobiology* **2017**, *36*, 413–425. [CrossRef] [PubMed]
44. Hobson, S.; Bateman, S.; Coe, J.B.; Oblak, M.L. Shelter Worker's Experiences and Perceptions of Capacity for Care (C4C). *J. Appl. Anim. Welf. Sci.* **2023**, *26*, 39–51. [CrossRef] [PubMed]
45. Rollin, B.E. Animal Rights as a Mainstream Phenomenon. *Animals* **2011**, *1*, 102–115. [CrossRef] [PubMed]
46. Mol, A.; Moser, I.; Pols, J. Care: Putting practice into theory. In *Care in Practice: On Tinkering in Clinics, Homes and Farm*; Mol, A., Moser, I., Pols, J., Eds.; Transcript: Bielefeld, Germany, 2010; pp. 7–26, ISBN 978-3-8376-1447-3.

veterinary sciences

MDPI

Article

Finite Element Analysis of Stress Distribution in Canine Lumbar Fractures with Different Pedicle Screw Insertion Angles

Ziyao Zhou [1,†], Xiaogang Shi [2,†], Jiahui Peng [1,†], Xiaoxiao Zhou [3], Liuqing Yang [4], Zhijun Zhong [1], Haifeng Liu [1], Guangneng Peng [1], Chengli Zheng [4,*] and Ming Zhang [5,*]

1 Teaching Veterinary Hospital, College of Veterinary Medicine, Sichuan Agricultural University, Chengdu 611130, China
2 Sichuan Wolong National Natural Reserve Administration Bureau, Wenchuan 623006, China
3 Chengdu Center for Animal Disease Prevention and Control, Chengdu 610041, China
4 Sichuan Institute of Musk Deer Breeding, Sichuan Institute for Drug Control, Chengdu 611130, China
5 College of Animal Science, Sichuan Agricultural University, Chengdu 611130, China
* Correspondence: zhengcl@scidc.org.cn (C.Z.); zhangming@sicau.edu.cn (M.Z.)
† These authors contributed equally to this work.

Simple Summary: Back injuries, especially fractures in the lower spine, are common in dogs. Surgeons often use special screws to stabilize these fractures, but the best angle to place these screws is not well understood. In this study, we used computer simulations to test how different screw angles affect stress and movement in a dog's spine. We created a 3D model of a beagle's lower back from CT scans and tested screws inserted at angles between 45° and 65°. Our results showed that screws placed at 58° caused the least stress on the spine, while angles between 56° and 60° provided the most stable fixation. This information can help veterinarians perform safer and more effective surgeries for dogs with spinal fractures. Future research should test these findings in real surgeries and different dog breeds.

Abstract: Pedicle screw fixation is a critical technique for stabilizing lumbar fractures in canines, yet the biomechanical implications of insertion angles remain underexplored. This study aims to identify optimal screw trajectories by analyzing stress distribution and deformation patterns in beagle lumbar segments (L6-L7) using finite element analysis (FEA). A 3D finite element model was reconstructed from CT scans of a healthy beagle, incorporating cortical/cancellous bone, intervertebral disks, and cartilage. Pedicle screws (2.4 mm diameter, 22 mm length) were virtually implanted at angles ranging from 45° to 65°. A 10 N vertical load simulated standing conditions. Equivalent stress and total deformation were evaluated under static loading. The equivalent stress occurred at screw–rod junctions, with maxima at 50° (11.73 MPa) and minima at 58° (3.25 MPa). Total deformation ranged from 0.0033 to 0.0064 mm, with the highest at 55° and the lowest at 54°. The 58° insertion angle demonstrated optimal biomechanical stability with minimal stress concentration, with 56–60° as a biomechanically favorable range for pedicle screw fixation in canine lumbar fractures, balancing stress distribution and deformation control. Future studies should validate these findings in multi-level models and clinical settings.

Keywords: canine lumbar spine; finite element analysis; pedicle screw fixation; insertion angle; biomechanical stability

1. Introduction

Common lumbar spinal disorders in canines encompass a variety of conditions including intervertebral disk herniation, vertebral fractures, neoplastic lesions affecting the spinal cord and nerve roots, and acute spinal cord injuries [1,2]. Among these pathologies, vertebral fractures or dislocations of traumatic etiology account for approximately 7% of cases, with lumbosacral region injuries being particularly prevalent [2]. In addition, when genetic or congenital ligament structural instability weakens the support function of the spine, potential pathological processes may lead to impaired bone integrity, which in turn can cause pathological fractures [3]. These lumbosacral fractures and dislocations constitute 39% of all spinal pathology cases documented in veterinary practice [4], which may lead to motor dysfunction, and urinary and fecal incontinence, seriously affecting quality of life in the animal.

Surgical intervention for lumbar fractures in canines is a method for achieving an anatomical reduction in the fractured vertebral segment, coupled with spinal cord decompression, contingent upon the neurological status and the extent of the traumatized region. Research has shown that there is a significant correlation between the treatment effect of canine spinal fractures and anatomical features [5]. In contrast to human spinal anatomy, canine vertebral dislocations and diverse fracture patterns typically present in recumbency, necessitating that pedicle screws and vertebral body implants be strategically positioned to traverse the maximal transverse diameter of the vertebral body [6]. This anatomical consideration enables secure fixation across adjacent vertebrae. The L6 segment is particularly susceptible to lumbar fractures in dogs, whereas the L7 segment has been identified as the most favorable site for pedicle screw fixation.

Finite element analysis (FEA) is a digital calculation method that uses computer-aided design to construct research models, simulate mechanical processes, and conduct mechanical analysis. The basic idea of FEA is to divide a certain structure into a number of small elements, assigning various mechanical properties (such as the elastic modulus and Poisson's ratio), to establish a matrix equation under loaded conditions for calculating the stress–strain relationship influenced by external forces, pressure, and other factors [7]. In the field of biomechanics, FEA can effectively simulate the mechanical response under complex physiological conditions, being especially suitable for the mechanical performance evaluation of the skeletal system [8]. Moreover, computer simulation can effectively reduce the use of experimental animals, in line with the 3R principle [9]. In human medicine, three-dimensional finite element analysis has been employed to enhance the accuracy of surgical approach evaluation and intraoperative planning in spinal surgery. Within the neurosurgical field, drill guide template technology has been developed for precise guidance during spinal fixation screw placement [10]. Previous studies have proposed a pedicle screw insertion angle range of 45° to 65° for canine lumbar fixation [11]; however, the specific optimal angles within this range for canine clinical applications remain unexplored.

Therefore, this study, using beagle dogs as experimental subject, initiated a three-dimensional (3D) model of the L6 and L7 lumbar segments in beagle dogs, derived from CT scan data. Subsequently, a finite element fracture model was constructed using Mimics, Geomagic, and Solidworks software. This model was then imported into Ansys for FEA to ascertain the distribution of equivalent stress and total deformation when pedicle screws were inserted at angles ranging from 45° to 65°. This analysis aims to provide a theoretical foundation for the biomechanical investigation of lumbar fractures in canines.

2. Materials and Methods

2.1. Ethics Statement

This study was approval by the Sichuan Agricultural University Animal Ethical and Welfare Committee with the approval number 20240221. All the animal procedures were conducted in strict accordance with the National Standards for Laboratory Animal Welfare (GB/T 35892-2018, China) [12], designed to minimize discomfort, including standardized anesthesia monitoring and post-procedural care.

2.2. CT Imagine Collection of Beagle Dog

A 2-year-old male beagle dog (body weight: 14.2 kg) was selected for this study, provided by the Sichuan Science and Technology Resources Sharing Platform of Beagle Dog Breeding and Experimental Technology Service. Anesthesia was induced with an intravenous injection of Zoletil® 50 (tiletamine-zolazepam, Virbac, France) at a dose of 10 mg/kg, and maintained with isoflurane (1–2% in oxygen) to ensure stable vital signs (heart rate: 80–100 bpm; respiratory rate: 15–20 breaths/min; rectal temperature: 38.0–39.0 °C) during scanning.

The dog was positioned in the supine position for CT scanning in Chengdu Ultrasound Imaging Center Co., Ltd., Chengdu, China (model uCT503e, featuring 40 rows and 40 layers). The CT scan settings were as follows: tube voltage, 120 kV; tube current, 250 mA; slice thickness, 3 mm; rotation time, 1.00 s; pitch, 0.6250; reconstruction matrix, 512 × 512; reconstruction slice thickness, 1 mm; bone window width, 2600 HU; window level, 800 HU; soft tissue window width, 300 HU; and window level, 40 HU. These settings were similar to those applied in our previous study [13].

The CT images were reviewed by a veterinary radiologist board-certified by China Veterinary Association to confirm the absence of spinal abnormalities (e.g., congenital deformities, fractures, or disk herniation) and ensure the clarity and completeness of L6-L7 segments.

2.3. Three-Dimensional Reconstruction of the Lumbar Spine

The 3D reconstruction tasks were performed on a personal computer equipped with an Intel(R) Core(TM) i5-8300H CPU operating at 2.30 GHz, with 16.0 GB of RAM, and running the Windows 10 Professional operating system, manufactured by ASUS Computer Inc. The 3D reconstruction and modeling were performed using Mimics 21.0, developed by Materialise in Vallda, Sweden. The DICOM format CT images were imported into Mimics 21.0 software for 3D reconstruction.

In Mimics 21.0, the threshold range for bone was set between 226 and 3071 HU, consistent with previous study [14]. After threshold selection, Mimics identified all regions within the lumbar spine. A base mask was created, and the "Measure" command was used to measure the length of the lumbar spine. Irrelevant regions (e.g., rib fragments, soft tissue) were manually erased using the "Edit masks" function, and voids within the vertebral body (due to partial volume effects) were filled with the "Close Holes" tool (hole diameter < 2 mm). To refine the segmentation of disconnected bony structures (e.g., articular processes), "Region growing" was applied with a seed point placed at the center of the L6 vertebral body, using a threshold tolerance of ±10 HU to ensure consistent inclusion of cancellous bone.

2.4. Smoothing and Repairing of the Bone Models

The 3D model was neatly computed using the "Compute Component" command. To ensure accuracy in finite element analysis, the model quality was set to "Optimal." Surface smoothing was performed using the "Remesh" operation in the 3-matic module of

Mimics. The "wrap" and "smooth" functions were applied, with the smooth coefficient set to 0.3 (balanced to preserve anatomical details while reducing noise). The model was then exported into STL format and imported into Geomagic Wrap 2017 (3D Systems, Rock Hill, SC, USA) for further smooth processing.

Cortical and cancellous bone were separated based on CT value thresholds (cortical bone: >800 HU; cancellous bone: 226–800 HU) and imported into SolidWorks 2021 (Dassault Systèmes, Vélizy-Villacoublay, France) for solidification. The L6 and L7 vertebrae were modeled as separate components and aligned at the global origin. An intervertebral disks (L6-L7) and articular cartilage were designed using anatomical measurements. The intervertebral disk and cartilage were created using sketching, copying, splitting, combining, and intersecting operations. The L6 segment was aligned vertically, and a fracture surface was cut perpendicular to the long axis of the vertebral body.

2.5. Assembly of the Lumbar Fracture Internal Fixation System

Pedicle screws were selected for internal fixation at different insertion angles. The screws were designed based on clinical standards and the size of the beagle. Four pedicle screws were modeled in Solidworks, each with a cap diameter of 4 mm, screw diameter of 2.4 mm, and total length of 22 mm [15]. The screws were inserted into the lumbar model at angles ranging from 45° to 47°, 50°–60° with 1° intervals, 63°, and 65°.

The assembled model was imported into Ansys, and each component was set as a separate part. Surface mesh inspection was performed to remove and manually repair deformed, overlapping, or disconnected meshes.

2.6. Finite Element Parameter Assignment and Loading

The cortical bone, cancellous bone, annulus fibrosus, nucleus pulposus, and ligaments were treated as linear elastic materials. The material properties were simplified to isotropic and linear elastic to improve computational efficiency. The pedicle screws and rods were made of titanium alloy (Ti-6Al-4V), with material parameters derived from previous studies. The material parameters are listed in Table 1.

Table 1. Material parameters of model components.

Material	Elastic Modulus (MPa)	Poisson's Ratio
Cortical Bone	12,000	0.30
Cancellous Bone	132	0.20
Nucleus Pulposus	1	0.49
Annulus Fibrosus	4.2	0.45
Articular Cartilage	11	0.40
Endplate	23.8	0.40
Ti-6Al-4V (Pedicle Screw)	110,000	0.30
Ti-6Al-4V (Fixation Rod)	110,000	0.30

2.7. Finite Element Analysis

For this study, a vertical load of 10 N was applied to simulate normal standing or walking conditions. The contact between the lumbar spine and implants was defined as fixed, with no sliding allowed. The intervertebral disk and vertebral body contacts were set as bonded, and the facet joints were defined as non-separable in the vertical direction but free to move horizontally. A 10 N force was applied to the spinous process, and the distal end of the lumbar spine was set as a fixed support.

The 3D finite element model was imported into Ansys. The equivalent stress and total deformation were calculated to compare the stress distribution and deformation in the lumbar spine with pedicle screws inserted at different angles under a 10 N vertical load.

3. Results

3.1. Model Establishment Results

A high-fidelity 3D model of the beagle dog's L6-L7 lumbar segments was successfully reconstructed from CT scan data, with clear visualization of anatomical details including the vertebral bodies, pedicles, transverse processes, spinous processes, intervertebral disk (L6-L7), and articular cartilage (Figure 1).

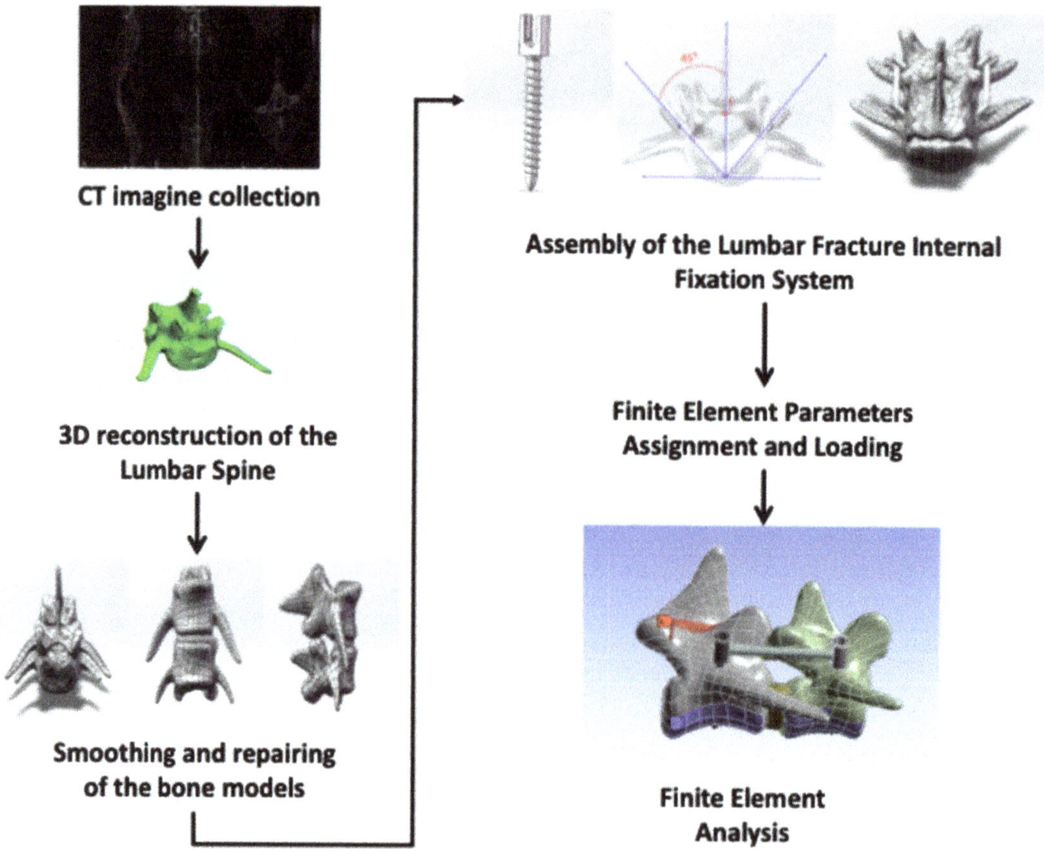

Figure 1. Scheme of 3D model and finite element analysis.

The pedicle screw model was designed with a cap diameter of 4 mm, a screw diameter of 2.4 mm, and a total length of 22 mm, featuring a threaded arc length of 106.56 mm. To systematically evaluate the influence of insertion angles on biomechanics, 15 finite element models were constructed, each with pedicle screws implanted at distinct angles relative to the vertebral sagittal plane: 45°, 47°, 50–60° (at 1° intervals), 63°, and 65°. Mesh quality assessment for each model (Table 2) showed node counts ranging from 161,757 to 168,027 and element counts ranging from 83,010 to 87,710, indicating high mesh quality suitable for subsequent FEA.

Table 2. The number of division nodes and elements in the pedicle screw implantation model under different angles.

Angles	Nodes	Elements	Angles	Nodes	Elements
45°	166,209	87,002	56°	168,027	87,710
47°	165,931	86,450	57°	165,850	86,478
50°	164,932	86,269	58°	164,115	85,784
51°	164,923	86,269	59°	165,859	86,478
52°	164,923	86,269	60°	166,389	86,852
53°	166,622	86,463	63°	161,757	83,027
54°	161,725	83,010	65°	166,622	86,436
55°	168,027	87,710			

3.2. Equivalent Stress at Different Pedicle Screw Implant Angles

Under a 10 N vertical load, the equivalent stress (von Mises) distribution across the L6-L7 fixation system varied significantly with pedicle screw insertion angles (Figure 2A). Consistent with biomechanical principles, stress concentration was most prominent at two key locations, (1) the screw neck (the junction between the screw cap and shaft) and (2) the interface between the screw cap and the fixation rod, which are recognized as potential high-risk regions for implant fatigue failure in spinal fixation systems.

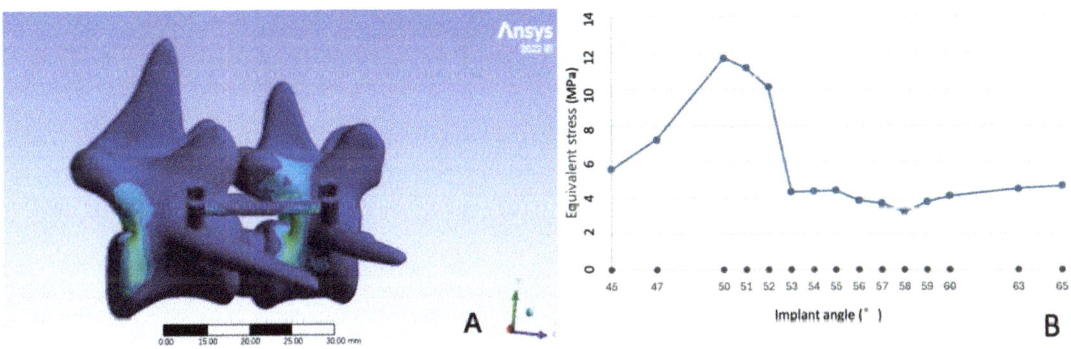

Figure 2. Equivalent stress at different pedicle screw insertion angles. (**A**) Distribution map of equivalent stress at 45°implant angle; (**B**) equivalent stress at different implant angles.

Quantitative analysis revealed that peak equivalent stress across all angles ranged from 3.2454 MPa to 11.7300 MPa (Figure 2B), with the highest value observed at 50° (11.7300 MPa) and the lowest at 58° (3.2454 MPa). Notably, stress values fluctuated substantially between 45° and 55° (range: 7.82–11.73 MPa), whereas angles between 56° and 60° showed relative stability (3.25–5.12 MPa).

3.3. Total Deformation at Different Pedicle Screw Insertion Angles

Total deformation (cumulative displacement) of the L6-L7 fixation system under 10 N loading was primarily localized to the intervertebral disk (L6-L7) and articular cartilage of the facet joints, consistent with their role as compliant structures in the spinal column (Figure 3).

Quantitative analysis of peak total deformation (Table 3) showed a range of 0.0033 mm to 0.0064 mm across all angles. The highest deformation was recorded at 55° (0.0064 mm), primarily concentrated in the posterior annulus fibrosus of the L6-L7 disk, suggesting increased tensile stress on the disk at this angle. Conversely, the lowest deformation was observed at 54° (0.0033 mm), with minimal displacement in both bony and soft tissue structures, indicating enhanced stability of the fixation system. Among them, angles between 51° and 54° consistently exhibited low deformation (<0.0040 mm), while angles at the extremes of the tested range (45°, 55°) showed higher displacement (>0.0055 mm), highlighting a potential relationship between insertion angle and system stability.

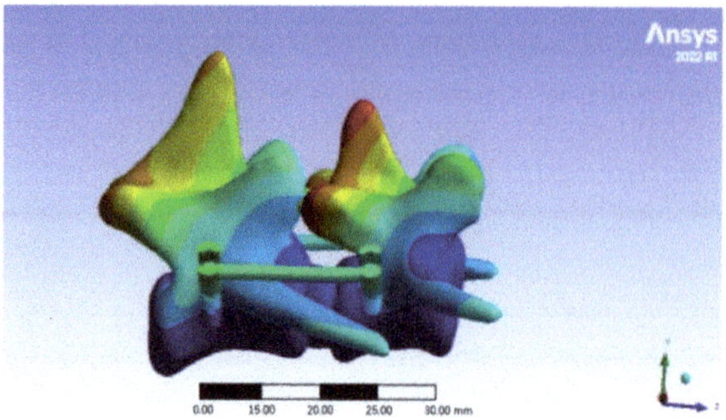

Figure 3. Total deformation distribution diagram of pedicle screw implantation in the lumbar spine (45°).

Table 3. Maximum total deformation values of pedicle screw implants at different implantation angles.

Angles	Maximum Total Deformation (mm)	Angles	Maximum Total Deformation (mm)
45°	0.0066	56°	0.0043
47°	0.0055	57°	0.0035
50°	0.0046	58°	0.0051
51°	0.0034	59°	0.0041
52°	0.0041	60°	0.0049
53°	0.0045	63°	0.0035
54°	0.0033	65°	0.0038
55°	0.0064		

4. Discussion

Finite element analysis has unique advantages in orthopedic biomechanics, including the following: (1) non-invasive modeling, which allows simulation of complex biomechanical behaviors without the need for physical specimens or animal trials; (2) stress–strain analysis that can predict stress distribution in bones, implants, and soft tissues, helping identify potential failure points; and (3) pre-surgical planning and virtual testing that helps surgeons evaluate different fixation strategies as well as implant performance under extreme conditions before real-world application [16,17]. Some canine lumbar biomechanical studies have existed. Yuki Kikuchi et al. [18] developed L1-L2 finite element models from canine CT data, applying 0.2–2 Nm incremental loads to demonstrate functional spinal unit stabilization. Their parallel mechanical testing and FEA of L1-L2 and L5-L6 segments showed strong correlation between computational predictions and experimental results. Tailong Yu et al. [19] designed a biodegradable dynamic stabilization system for interfaced lumbar fusion, demonstrating comparable fusion outcomes and biomechanical performance to those of rigid fixation with safe in vivo degradation. However, our study is the first study using the FEA method for determining the best angle for pedicle screw insertion in L6-L7.

During normal standing or walking, dogs primarily experience vertical downward pressure on the lumbar spine due to their weight being transmitted through the limbs. TAE-HONG LIM et al. [20] conducted a comprehensive biomechanical evaluation of the intact L6-L7 functional spinal unit. Their model predicted key parameters including flexion angle, axial stiffness, and facet joint contact forces. The study established that applying a load equivalent to 7% of canine body weight represents a physiologically reasonable loading condition for lumbar spine analysis. Given that the present experiment primarily investigates stress distribution during normal walking and standing states rather than overload conditions, the vertical loading force was set at an average of 10 N for beagles (7% of 14 kg), with the direction oriented vertically downward. A frictional contact interaction was defined between the lumbar vertebrae and the implant, which was modeled as a fixed interface.

In FEA, equivalent stress characterizes the internal resistance of a material to deformation under external loading. The displacement reflects the relative motion between the implant and bone tissue, serving as a critical indicator for evaluating the stability of the internal fixation system. Total deformation value refers to the microscopic displacement of bone tissue under mechanical loading at fracture sites, which serves as an indicator for assessing bone stability [21]. A study [22] reported that titanium alloy implants undergo irreversible deformation at a stress of 450 MPa, with fracture risks exceeding 600 MPa. Our results demonstrated that the greatest implant stress was 11.73 MPa, with fracture risks being much lower. Comparative analysis of maximum equivalent stress nephograms across internal fixation models revealed consistent stress distribution patterns in pedicle screws, with maximum axial displacement occurring proximally at loading sites and gradually decreasing distally. The 51°–54° and 59°–65° insertion angles showed lower displacement magnitudes compared to overall values, with peak total deformation ranging between 0.0033 and 0.0064 mm. These findings illustrated that the degree of postoperative micromotion of the pedicle screw fixation end is within an acceptable range, reducing the risk of surgical failure caused by displacement [23].

Although the FEA results are all within the acceptable range of lumbar spine biomechanics, notable abrupt deformations or stress fluctuations were observed in response to specific angular deviations. For example, the maximum intervertebral disk deformation (0.0064 mm) occurred at 55°, while the minimum deformation (0.0033 mm) was observed at 54°; the stress at the L6 fracture site reached 8.7 MPa at 50°, but decreased to 3.5 MPa at 58°. Two potential mechanistic explanations may underlie these abrupt mechanical changes: (1) Anatomical influence of the lumbar spine: lumbar vertebral bodies exhibit an irregular elliptical geometry. During screw implantation, as the trajectory approaches the elliptical apices of the vertebral body, a mechanical threshold is reached, triggering dramatic stress–deformation responses. (2) Bone tissue transition with angular variation: alterations in screw angulation led to a shift in contact from cancellous to cortical bone within the transverse process, directly modulating mechanical loading profiles and stress distribution patterns. Although vertebral displacement represents a primary determinant of crack initiation and propagation prior to fracture, minor deformations may be fixed by Wolff's law [24]. During screw insertion, cancellous bone within pedicles becomes compacted around screws, creating cavities. Following fracture healing and implant removal, trabecular bone within these cavities undergoes thickening under mechanical stimulation, leading to displacement variations as trabeculae reorganize into denser, more regular patterns [25]. Additionally, there are varying ranges of macroscopic yield strain before bone failure, while bone strain also functions as a mechanical signal triggering bone remodeling during healing processes [26,27]. Therefore, postoperative rehabilitation should include appropriate activity selection, as controlled lumbar displacement may facilitate fracture healing [28].

Our results indicate that optimal screw trajectories in beagles target high-density regions at the ventrobasal pedicle–vertebral junction. For lumbar vertebral fractures, bilateral pedicle screw fixation systems can significantly suppress abnormal intervertebral motion, and this rigid fixation method helps maintain postoperative spinal mechanical stability. While L7 permits similar implantation to that permitted by L1-L6, iliac wing anatomy significantly increases procedural difficulty—a common challenge across canine models. However, bilateral pedicle screw fixation requires extensive dissection of the paraspinal muscles, which may lead to denervation atrophy and postoperative lumbar weakness syndrome. Additionally, compared with unilateral fixation, bilateral pedicle screw fixation carries a higher risk of nerve root injury [29]. However, the results of this study showed that larger angles were associated with lower strain and stress, which might be attributed to the fact that partial pedicle screws enter the cortical bone, leading to harder contact. Therefore, this could increase the risk of Grade 1 or 2 canal breach in clinical practice. Consequently, clinics should incorporate anatomical considerations to identify optimal screw trajectories while minimizing neural interference. Proper screw placement results in acceptable postoperative micromotion at fixation termini, thereby reducing displacement-related failure risks [23]. FEA can simulate pedicle screw placement on the side of nerve root decompression, theoretically reducing nerve injury by 50% after screw insertion [30].

Frankly, our study has several limitations. For example, static loading conditions may not fully reflect the complex mechanical environment of the canine lumbar spine during daily activities. Our model applied a constant 10 N vertical load to simulate standing or slow walking, but dogs experience dynamic loads that fluctuate with movement—for example, peak vertical forces can reach two to three times the body weight during running or jumping [31]. Additionally, breed-specific anatomical variability remains a critical consideration. We selected beagles as a representative medium-sized breed to minimize size-related bias; their L7 pedicles (>3 mm width) provide adequate implantation space, favoring transpedicular fixation, while lumbar spine morphology varies substantially across canine breeds. For example, large breeds (e.g., German Shepherds) exhibit longer vertebral bodies and thicker pedicles (4.5–5.0 mm in diameter) compared to small breeds (e.g., Yorkshire Terriers, pedicle diameter: 2.0–2.5 mm). These differences could affect optimal screw trajectory, as a 58° angle—identified as biomechanically favorable in beagles—might lead to pedicle breach in smaller breeds or inadequate purchase in larger breeds. Breeds with narrower L7 pedicles (e.g., Bulldogs) may also require alternative angles or screw sizes. More importantly, the model did not account for patient-specific factors such as bone quality (e.g., osteoporosis, which is common in elderly dogs), which can drastically reduce screw pullout strength and alter stress distribution [32].

5. Conclusions

In conclusion, the results of this study show that the peak stress fluctuates significantly between 45° and 55°, with the highest stress occurring at 50°. The stress is more stable between 56° and 60°, with the lowest stress at 58°, indicating the best biomechanical stability. Finite element fracture models developed using spinal loading scoring systems demonstrate superior clinical relevance, with simulation results showing strong concordance with biomechanical characteristics from actual cases [33]. This approach enables the quantitative assessment of vertebral injury severity, fracture fragment distribution, displacement magnitude, and kyphotic correction, all of which can be effectively simulated through finite element methods [34]. However, finite element modeling software presents considerable operational challenges, demanding substantial computational expertise. Integration of FEA with laser and/or robot-assisted guidance, or 3D printing technology, which is emerging in the veterinary field, e.g., 3D-printed personalized guidance devices capable of converting medical imaging data into anatomically accurate models, may enhance preoperative planning and surgical success rates in lumbar procedures. Future studies should systematically evaluate breed-specific implantation parameters.

Author Contributions: Conceptualization, Z.Z. (Ziyao Zhou) and J.P.; methodology, J.P.; software, X.S.; formal analysis, X.S.; data curation, X.S., X.Z. and L.Y.; writing—original draft preparation, Z.Z. (Ziyao Zhou) and J.P.; writing—review and editing, X.S., G.P., C.Z. and M.Z.; visualization, Z.Z. (Zhijun Zhong) and H.L.; supervision, C.Z. and M.Z.; funding acquisition, C.Z. All authors have read and agreed to the published version of the manuscript.

Funding: This research was funded by the Science and Technology program of the Sichuan Medical Products Administration (2024010).

Institutional Review Board Statement: This study was approved by the Sichuan Agricultural University Animal Ethical and Welfare Committee (20240221).

Informed Consent Statement: Not applicable.

Data Availability Statement: The original contributions presented in the study are included in the article material, and further inquiries can be directed to the corresponding authors.

Acknowledgments: We are grateful for the support of Sichuan Science and Technology Resources Sharing Platform of Beagle Dog Breeding and Experimental Technology Service.

Conflicts of Interest: The authors declare no conflict of interest.

References

1. Della Valle, G.; Di Dona, F.; Mennonna, G.; Lamagna, B.; Pasolini, M.; Caterino, C.; Lamagna, F.; Fatone, G. Traumatic lumbosacral joint dislocation in 3 dogs: Clinical presentation, diagnosis, treatment and short-term follow-up. *Pak. Vet. J.* **2021**, *41*, 97–101.
2. Jeffery, N.D. Vertebral fracture and luxation in small animals. *Vet. Clin. N. Am. Small Anim. Pract.* **2010**, *40*, 809–828. [CrossRef] [PubMed]
3. Steinberg, H.S.; Coates, J.R. *Diagnosis of and Treatment Options for Disorders of the Spine*; Wiley: Hoboken, NJ, USA, 2013.
4. Orgonikova, I.; Brocal, J.; Cherubini, G.; Palus, V. Vertebral fractures and luxations in dogs and cats, part 1: Evaluation of diagnosis and prognosis. *Companion Anim.* **2021**, *26*, 1–10. [CrossRef]
5. Gallastegui, A.; Davies, E.; Zwingenberger, A.; Nykamp, S.; Rishniw, M.; Johnson, P. MRI has limited agreement with CT in the evaluation of vertebral fractures of the canine trauma patient. *Vet. Radiol. Ultrasound* **2019**, *60*, 533–542. [CrossRef] [PubMed]
6. Bali, M.S.; Lang, J.; Jaggy, A.; Spreng, D.; Doherr, M.G.; Forterre, F. Comparative study of vertebral fractures and luxations in dogs and cats. *Vet. Comp. Orthop. Traumatol.* **2009**, *22*, 47–53. [PubMed]
7. Liu, R.; Yang, J.; Zhu, Y.; Zhou, X.; Zhou, Q.; Liang, T.; Wang, H.; Luo, Y.; Xie, Y.; Liu, H. A forecasting model for suitable dental implantation in canine mandibular premolar region based on finite element analysis. *BMC Vet. Res.* **2024**, *20*, 353. [CrossRef] [PubMed]
8. Shi, D.; Wang, F.; Wang, D.; Li, X.; Wang, Q. 3-D finite element analysis of the influence of synovial condition in sacroiliac joint on the load transmission in human pelvic system. *Med. Eng. Phys.* **2014**, *36*, 745–753. [CrossRef] [PubMed]
9. Long, H.; Zhang, H.; Deng, L.; Li, C.; Zhu, Y.; Ling, S.; Li, D.; Zhou, X.; Liu, H.; Zhong, Z. Biomechanics of Dental Implantation in the Giant Panda (Ailuropoda melanoleuca): A Comparative Study Using Finite Element Analysis. *J. Vet. Dent.* **2024**, *42*, 273–280. [CrossRef] [PubMed]
10. Pu, X.; Yin, M.; Ma, J.; Liu, Y.; Chen, G.; Huang, Q.; Zhao, G.; Lu, T.; Yao, S.; Chen, Q.; et al. Design and Application of a Novel Patient-Specific Three-Dimensional Printed Drill Navigational Guiding in Atlantoaxial Pedicle Screw Placement. *World Neurosurg.* **2018**, *114*, e1–e10. [CrossRef] [PubMed]
11. Kim, J.-Y.; Kwak, H.-H.; Woo, H.-M.; Kim, J. Enhancing the Accuracy of Pedicle Screw Placement Using 3D-Printed Screw-Guiding Techniques in the Lumbosacral Region for Small Breed Dogs: A Cadaveric Study. *Animals* **2024**, *15*, 14. [CrossRef] [PubMed]
12. MacArthur, C.J.A.; Sun, D. Guidelines for the ethical review of laboratory animal welfare People's Republic of China National Standard GB/T 35892-2018. *Anim. Models Exp. Med.* **2020**, *3*, 103–113.

13. Peng, J.; Yang, J.; Liu, R.; Liu, H.; Zhong, Z.; Peng, G.; Zhang, K.; Zheng, C.; Zhang, M.; Zhou, Z. Evaluating the concept of three-dimensional printing guided endodontics in the dog. *Front. Vet. Sci.* **2024**, *11*, 1481612. [CrossRef] [PubMed]
14. Ford, J.M.; Decker, S.J. Computed tomography slice thickness and its effects on three-dimensional reconstruction of anatomical structures. *J. Forensic Radiol. Imaging* **2016**, *4*, 43–46. [CrossRef]
15. Kim, Y.Y.; Choi, W.S.; Rhyu, K.W. Assessment of pedicle screw pullout strength based on various screw designs and bone densities-an ex vivo biomechanical study. *Spine J. Off. J. N. Am. Spine Soc.* **2012**, *12*, 164–168. [CrossRef] [PubMed]
16. Yang, K. *Basic Finite Element Method as Applied to Injury Biomechanics*; Elsevier: Amsterdam, The Netherlands, 2017; pp. 1–730.
17. Lewis, G.S.; Mischler, D.; Wee, H.; Reid, J.S.; Varga, P. Finite Element Analysis of Fracture Fixation. *Curr. Osteoporos. Rep.* **2021**, *19*, 403–416. [CrossRef] [PubMed]
18. Kikuchi, Y.; Shimada, M.; Takahashi, F.; Yamaguchi, S.; Hara, Y. Finite element analysis shows minimal stability difference between individualized mini-hemilaminectomy-corpectomy and partial lateral corpectomy in a dog model. *Am. J. Vet. Res.* **2024**, *85*, 12. [CrossRef] [PubMed]
19. Yu, T.; Zheng, L.; Chen, G.; Wang, X.; Chi, H.; Song, C.; Xi, C.; Yan, J. A novel dynamic fixation system with biodegradable components on lumbar fusion between articular processes in a canine model. *J. Eng. Med.* **2020**, *234*, 738–748. [CrossRef] [PubMed]
20. Lim, T.-H.; Goel, V.K.; Weinstein, J.N.; Kong, W. Stress analysis of a canine spinal motion segment using the finite element technique. *J. Biomech.* **1994**, *27*, 1259–1269. [CrossRef] [PubMed]
21. Atsumi, M.; Park, S.H.; Wang, H.L. Methods used to assess implant stability: Current status. *Int. J. Oral Maxillofac. Implant.* **2007**, *22*, 743–754.
22. Bahl, S.; Suwas, S.; Chatterjee, K. Comprehensive review on alloy design, processing, and performance of β Titanium alloys as biomedical materials. *Int. Mater. Rev.* **2020**, *66*, 114–139. [CrossRef]
23. Li, W.; Zhang, P.; Gan, F. Biomechanical characterization of bilateral pedicle screw internal fixation combinations on lumbar vertebrae. *Acta Bioeng. Biomech.* **2023**, *25*, 43–51. [CrossRef]
24. Giorgio, I.; dell'Isola, F.; Andreaus, U.; Misra, A. An orthotropic continuum model with substructure evolution for describing bone remodeling: An interpretation of the primary mechanism behind Wolff's law. *Biomech. Model. Mechanobiol.* **2023**, *22*, 2135–2152. [CrossRef] [PubMed]
25. Lin, C.Y.; Kang, J.H. Mechanical Properties of Compact Bone Defined by the Stress-Strain Curve Measured Using Uniaxial Tensile Test: A Concise Review and Practical Guide. *Materials* **2021**, *14*, 4224. [CrossRef] [PubMed]
26. Oryan, A.; Monazzah, S.; Bigham-Sadegh, A. Bone injury and fracture healing biology. *Biomed. Environ. Sci.* **2015**, *28*, 57–71. [CrossRef] [PubMed]
27. Hoenig, T.; Ackerman, K.E.; Beck, B.R.; Bouxsein, M.L.; Burr, D.B.; Hollander, K.; Popp, K.L.; Rolvien, T.; Tenforde, A.S.; Warden, S.J. Bone stress injuries. *Nat. Rev. Dis. Primers* **2022**, *8*, 26. [CrossRef] [PubMed]
28. Meng, B.; Bunch, J.; Burton, D.; Wang, J. Lumbar interbody fusion: Recent advances in surgical techniques and bone healing strategies. *Eur. Spine J.* **2021**, *30*, 22–33. [CrossRef] [PubMed]
29. Liu, H.; Xu, Y.; Yang, S.D.; Wang, T.; Wang, H.; Liu, F.Y.; Ding, W.Y. Unilateral versus bilateral pedicle screw fixation with posterior lumbar interbody fusion for lumbar degenerative diseases: A meta-analysis. *Medicine* **2017**, *96*, e6882. [CrossRef] [PubMed]
30. Dong, J.; Rong, L.; Feng, F.; Liu, B.; Xu, Y.; Wang, Q.; Chen, R.; Xie, P. Unilateral pedicle screw fixation through a tubular retractor via the Wiltse approach compared with conventional bilateral pedicle screw fixation for single-segment degenerative lumbar instability: A prospective randomized study. *J. Neurosurg. Spine* **2014**, *20*, 53–59. [CrossRef] [PubMed]
31. Voss, K.; Galeandro, L.; Wiestner, T.; Haessig, M.; Montavon, P.M. Relationships of body weight, body size, subject velocity, and vertical ground reaction forces in trotting dogs. *Vet. Surg.* **2010**, *39*, 863–869. [CrossRef] [PubMed]
32. Shea, T.M.; Laun, J.; Gonzalez-Blohm, S.A.; Doulgeris, J.J.; Lee III, W.E.; Aghayev, K.; Vrionis, F.D. Designs and techniques that improve the pullout strength of pedicle screws in osteoporotic vertebrae: Current status. *BioMed Res. Int.* **2014**, *2014*, 748393. [CrossRef] [PubMed]
33. Colombo, C.; Libonati, F.; Rinaudo, L.; Bellazzi, M.; Ulivieri, F.M.; Vergani, L. A new finite element based parameter to predict bone fracture. *PLoS ONE* **2019**, *14*, e0225905. [CrossRef] [PubMed]
34. Van den Munckhof, S.; Zadpoor, A.A. How accurately can we predict the fracture load of the proximal femur using finite element models? *Clin. Biomech.* **2014**, *29*, 373–380. [CrossRef] [PubMed]

veterinary sciences

MDPI

Communication

Real-Time Monitoring of Cardiac Output Using Non-Invasive Impedance Cardiography in Dogs: A Pilot Study on Heartworm Extraction and Gastric Decompression

Daesik Kim [1], Seol-Gi Park [2] and Min Su Kim [1,*]

[1] Veterinary Emergency Medicine, Department of Veterinary Clinical Science, Research Institute for Veterinary Science, College of Veterinary Medicine, Seoul National University, Seoul 08826, Republic of Korea; daxic88@snu.ac.kr

[2] Incheon Sky Animal Hospital, Incheon 21555, Republic of Korea; roronanah@gmail.com

* Correspondence: minsukim@snu.ac.kr

Simple Summary: Non-invasive monitoring tools can significantly improve veterinary care for critically ill animals. This study investigated impedance cardiography (ICG) as a method to monitor heart function in dogs with two serious conditions: gastric dilatation and heartworm disease. We examined how key cardiovascular parameters changed before, during, and after standard treatments for these conditions. Our findings revealed that both heartworm extraction and gastric decompression led to measurable improvements in heart function, with cardiac index increasing by 14.71% and 28.53%, respectively. These results demonstrate that ICG can effectively detect hemodynamic changes without invasive procedures, offering veterinarians a practical tool for monitoring patient response to treatment. This approach could improve clinical decision-making and potentially lead to better outcomes for dogs with these critical conditions.

Abstract: Background: Gastric dilatation (GD) and heartworm disease (HW) can significantly impact cardiovascular function, necessitating timely intervention. This study evaluated hemodynamic changes in these conditions using impedance cardiography (ICG), a non-invasive technique for real-time monitoring. Methods: Cardiac parameters were measured using ICG in healthy controls, HW, and GD groups at baseline, during intervention, and post-treatment. Statistical analyses involved repeated measures ANOVA with post hoc comparisons, and relative changes (%) quantified hemodynamic improvements. Results: A significant increase in cardiac index (CI) occurred post-treatment in both HW (+14.71%, $p = 0.0102$) and GD (+28.53%, $p = 0.0336$) groups. Cardiac output (CO) exhibited an increasing trend, though without significance ($p > 0.05$). Mean arterial pressure (MAP) remained stable, although the GD group exhibited an upward trend post decompression ($p = 0.1213$). Conclusions: Heartworm extraction and gastric decompression induced measurable cardiovascular improvements, with distinct recovery patterns. Non-invasive monitoring effectively captured cardiovascular changes, suggesting its utility as an adjunctive tool in clinical assessment.

Keywords: gastric dilatation; heartworm; cardiac output; non-invasive monitoring; dog; interventional extraction

1. Introduction

Accurate and timely monitoring of cardiac output (CO) is essential in managing patients with critical cardiovascular conditions. CO, defined as the volume of blood ejected by the heart per minute, is a fundamental indicator of cardiac function and overall hemodynamic status [1,2]. Traditional methods for measuring CO, such as thermodilution, transesophageal echocardiography (TEE), arterial pressure waveform analysis (APWA), and CT/MRI-based evaluation, have specific advantages and limitations. Thermodilution is highly accurate and considered the gold standard [3,4]; however, it is invasive, technically demanding, and carries risks such as infection and bleeding [5]. TEE provides detailed cardiac imaging but is invasive and may induce discomfort or complications in patients with esophageal disorders [6]. APWA allows continuous real-time monitoring but can exhibit significant inaccuracies under specific hemodynamic conditions [7]. While CT and MRI deliver precise three-dimensional cardiac assessments, their high costs and inability to perform real-time monitoring limit practical use in emergency or acute care settings [8–10]. These limitations underscore the need for a non-invasive, reliable, and real-time CO monitoring technique, particularly in critical and rapidly evolving clinical situations.

Impedance cardiography (ICG) has emerged as a promising non-invasive modality for continuous cardiac output (CO) monitoring. It provides real-time hemodynamic data by detecting thoracic electrical impedance changes associated with each cardiac cycle, without requiring catheter insertion. Its simplicity and rapid deployment make it highly attractive for a variety of clinical scenarios [4,11]. Although its absolute accuracy can be influenced by factors such as thoracic conformation or fluid accumulation, ICG has consistently demonstrated reliable sensitivity in detecting directional changes in CO over time [12–14]. Direct comparisons between ICG and the gold standard thermodilution (TD) method have shown moderate agreement, with electrical cardiometry (a form of ICG) achieving a bias of −0.14 L/min and concordance rates exceeding 85% for trend detection [4,7]. Compared to other non-invasive methods like echocardiography and arterial pressure waveform analysis, ICG offers continuous real-time monitoring without operator dependency or the need for arterial catheterization [12,15–17]. Furthermore, a comprehensive evaluation of multiple CO monitoring techniques under hemodynamic stress confirmed that, while ICG's absolute accuracy remains modest, its favorable trend-tracking ability is clinically valuable [7]. It was originally developed for human medicine, but now it is increasingly used in veterinary medicine [16,17].

Heartworm (HW) disease significantly impacts cardiac output (CO) through increased pulmonary artery pressure, right ventricular overload, and tricuspid valve dysfunction, leading to impaired forward flow [18,19]. Similarly, gastric dilatation (GD) leads to CO reduction due to increased intra-abdominal pressure and impaired venous return [4,20]. Additionally, vagally mediated bradycardia further suppresses CO [21–23]. Although heartworm disease and gastric dilatation differ in their pathophysiological mechanisms—with the former involving progressive pulmonary hypertension and the latter characterized by acute mechanical compression of major vessels—both result in compromised cardiac output and require rapid intervention expected to elicit abrupt hemodynamic shifts. This study aimed to evaluate the feasibility and utility of non-invasive ICG for real-time CO monitoring in patients diagnosed with HW and GD.

By assessing CO variations during parasite removal in HW and decompression procedures in GD, this study sought to determine whether ICG can provide consistent trend-tracking data to support clinical decision-making in acute care settings, despite potential variations in absolute accuracy under different physiological conditions.

The aim of this study was to evaluate the feasibility and utility of non-invasive impedance cardiography (ICG) for real-time cardiac output monitoring in dogs with heartworm disease (HW) and gastric dilatation (GD). We hypothesized that ICG would reliably detect relative hemodynamic changes associated with therapeutic interventions (heartworm extraction and gastric decompression), thereby supporting its potential use as a clinical monitoring tool. This study specifically evaluated cardiac output (CO) along with associated parameters, including cardiac index (CI) and mean arterial pressure (MAP), to characterize hemodynamic changes before, during, and after intervention. Through this investigation, we explored whether ICG could provide a practical, efficient, and sensitive method for tracking cardiac function, particularly in conditions characterized by rapid fluctuations in CO. These findings may provide a basis for future studies and broader clinical evaluation of ICG in veterinary emergencies and critical care settings.

2. Material and Methods

This study was conducted in compliance with ethical guidelines for animal research and was approved by the Institutional Animal Care and Use Committee (IACUC) of Seoul National University (Approval No. SNU-200226-7-2, SNU-240115-9-1). A total of 16 dogs were included and categorized into three groups: six healthy controls, six HW cases, and four GD cases. The healthy control group was used to establish reference values for CO and mean arterial pressure (MAP), while the HW and GD groups underwent interventional procedures with ICG monitoring before and after treatment. The control group consisted of six healthy adult Beagle dogs (age range: 2–4 years; body weight: 5.5–10.0 kg). All animals were purpose-bred and had not been previously enrolled in any experimental procedures. Prior to inclusion in this study, each dog underwent a comprehensive health screening that included a thorough physical examination, complete blood count, serum biochemistry profile, thoracic radiography, and abdominal ultrasonography to rule out any underlying disease. Only dogs with no clinical abnormalities and normal laboratory and imaging findings were included as controls. All animals were housed in a controlled environment with standardized feeding and care protocols throughout the study period.

The HW and GD groups consisted of client-owned dogs that presented for therapeutic intervention (Figure 1). For the HW group, dogs included various breeds such as Poodle, Yorkshire Terrier mix, Pit Bull Terrier, and mixed-breed dogs (age range: 2–12 years; body weight: 3.72–46.0 kg). Patient selection was based on the American Heartworm Society (AHS) guideline criteria, specifically targeting dogs for whom surgical heartworm extraction was indicated due to significant worm burden or right atrial involvement as determined by echocardiography. All HW cases presented with moderate to severe heartworm disease, characterized by multiple adult worms visualized within the pulmonary arteries and/or right heart chambers. Mild clinical signs such as intermittent coughing, mild exercise intolerance, and mild ascites were observed in some cases. While none of the HW dogs in this study had developed caval syndrome at the time of intervention, we recognize the importance of including such cases to fully capture the severe end of the disease spectrum. All dogs exhibited mild to moderate pulmonary hypertension on echocardiography without evidence of severe right ventricular dysfunction or overt right-sided heart failure.

For the GD group, dogs included Dachshund, Samoyed, and Welsh Corgi breeds (age range: 1–8 years; body weight: 6.62–23.5 kg). Diagnosis of acute gastric dilatation was made based on physical examination, a detailed history, and confirmation with abdominal radiographs. Clinical signs included abdominal distension, non-productive retching, and varying degrees of cardiovascular compromise. Based on clinical assessment at presentation, one dog was classified as having severe cardiovascular compromise (poor pulse quality and prolonged capillary refill time), two dogs exhibited moderate compromise, and one showed mild cardiovascular effects. Emergency stabilization, including orogastric decompression and/or surgical intervention, was promptly performed. In all GD cases, ICG monitoring was conducted after initial stabilization to minimize the effects of acute hemodynamic instability on measurement reliability.

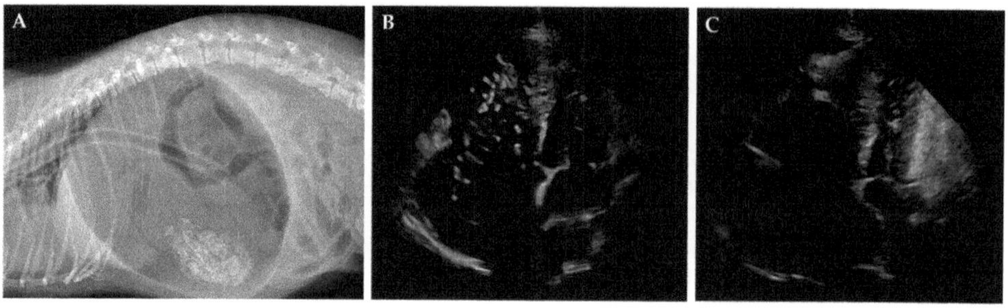

Figure 1. Radiographic and echocardiographic findings in cases of gastric dilatation and heartworm disease. (**A**) Lateral radiograph of a dog with severe gastric dilatation, showing a markedly distended, gas-filled stomach. An orogastric tube for decompression is visible. (**B**,**C**) Right parasternal short-axis echocardiographic images before (**B**) and after (**C**) heartworm removal. Heartworms appear as short, segmented, parallel echogenic structures (**B**), which are no longer visible post-extraction (**C**).

Informed consent for participation, including the use of non-invasive impedance cardiography (ICG) monitoring during therapeutic procedures, was obtained from all owners prior to enrollment.

2.1. Impedance Cardiography Setup

Impedance cardiography (ICG) is a non-invasive technique for continuous cardiac output (CO) and stroke volume (SV) monitoring. In this study, the PhysioFlow® system (version 1.0.7 RC9.15 SVV Edition Lab1 Enduro, Manatec Biomedical, Bristol, PA, USA) was used to obtain real-time hemodynamic data. This device utilizes Signal-Morphology Impedance Cardiography (SM-ICG™) technology, which analyzes the shape of the impedance waveform rather than relying on baseline thoracic impedance (Z0), allowing for more robust measurements even under variable clinical conditions. ICG operates by transmitting a low-magnitude, high-frequency alternating current through the thorax via surface electrodes. Thoracic impedance changes (ΔZ) during the cardiac cycle are detected and differentiated (dZ/dt), corresponding to aortic blood flow velocity. Six electrodes were applied according to a standardized configuration [12]: two were placed dorsally at the level of the 4th and 6th cervical vertebrae near the left jugular vein, one on the right side of the sternum at the 5th intercostal space, one on the left lateral thorax near the 13th rib, and two just below the xiphoid process (Figure 2). To optimize signal acquisition, electrodes were attached using clip-type leads with conductive gel after clipping and cleaning the skin with alcohol. Placement accuracy was maintained within ± 3 cm of anatomical landmarks, and each setup was verified by two operators to ensure consistency. The system was calibrated according to manufacturer specifications before each monitoring session. Only data with a signal quality $\geq$80% were included in the analysis, and measurements contaminated by motion artifacts or electrical interference were excluded. ICG measurement intervals were set at 5 s, and 15 repeated values were acquired per phase for each dog, allowing for both assessment of immediate changes and calculation of averaged values. For the GD group, CO and other hemodynamic parameters were measured in three distinct phases: before decompression (pre-decompression), immediately after decompression (post-decompression), and during the stabilization phase (post-stabilization). These measurements evaluated the hemodynamic instability caused by elevated intra-abdominal pressure and its subsequent relief. For the HW group, CO monitoring was conducted also in three phases: prior to the extraction procedure (pre-extraction), during HW extraction (extraction phase), and during the recovery phase after extraction (post-stabilization).

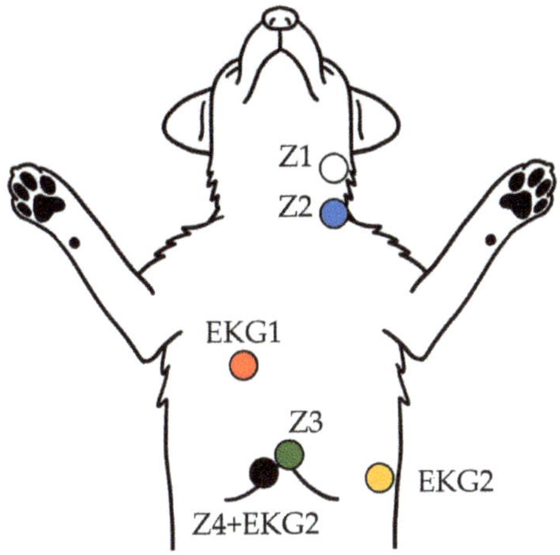

Figure 2. Schematic diagram of ICG and EKG electrode placement in dogs. The illustration shows the standardized surface electrode placement used to acquire thoracic impedance (Z1–Z4) and ECG (EKG1–2) signals for non-invasive cardiac output monitoring in dogs positioned in dorsal recumbency.

This consistent approach provided a comprehensive and accurate assessment of the hemodynamic effects during interventions in both the GD and HW groups. Both CO and CI values were derived from impedance measurements, with CI calculated as CO normalized to body surface area (BSA) and expressed in liters per minute per square meter ($L/min/m^2$).

2.2. Experimental Protocol

Prior to enrollment, all dogs underwent a physical examination and baseline diagnostic screening, including complete blood count, serum biochemistry, and thoracic radiography or ultrasonography when indicated, to ensure the absence of significant comorbidities unrelated to heartworm disease or gastric dilatation. GD cases were retrospectively confirmed to be free of other underlying conditions following full clinical workup during hospitalization. All dogs (NC, HW, and GD groups) received butorphanol (0.2–0.4 mg/kg IV) as premedication. To reduce procedural side effects, each dog was medicated starting from 3 days prior to the procedure and continuing up to the day of extraction. This pre-treatment was intended to minimize potential perioperative complications during heartworm removal. The medications administered included prednisolone (0.5 mg/kg PO BID), doxycycline monohydrate (5 mg/kg PO BID), silymarin (20 mg/kg PO SID), famotidine (0.5 mg/kg PO BID), clopidogrel hydrogen sulfate (2 mg/kg PO SID), cefaclor (20 mg/kg PO BID), and furosemide (1 mg/kg PO SID). In the NC and HW groups, general anesthesia was induced with alfaxalone (1–4 mg/kg IV, titrated to effect) and maintained with isoflurane (1–4%) following tracheal intubation. In the GD group, sedation was performed using butorphanol alone, with one dog undergoing endotracheal intubation and general anesthesia for surgical correction, and the remaining three managed without intubation during gastric decompression.

For the healthy control group (NC, $n = 6$), a standardized anesthesia protocol was implemented to evaluate ICG measurement stability in the absence of pathological conditions or interventional procedures. Following initial physical examination and confirmation of health status, these animals underwent general anesthesia maintained for a total duration of one hour. During this period, no specific interventions were performed, allowing assessment of hemodynamic parameter stability under controlled conditions.

CO and MAP measurements were obtained at three distinct time points: (1) stabilization phase (15–20 min after anesthesia induction, once cardiovascular parameters had stabilized), (2) mid-anesthesia phase (approximately 30 min into anesthesia), and (3) late anesthesia phase (approximately 50–60 min into anesthesia). This protocol was designed specifically to evaluate the consistency and reliability of ICG measurements over time in subjects with presumed stable hemodynamic status, thereby establishing both reference ranges and the trending capability of the monitoring system.

In the HW group, impedance cardiography (ICG) monitoring was performed across three defined phases: (1) pre-extraction phase: baseline cardiac output (CO) and mean arterial pressure (MAP) were recorded after anesthesia induction, jugular venotomy, and placement of the heartworm extraction device under fluoroscopic guidance. Measurements were obtained during a period of cardiovascular stability before initiating heartworm removal; (2) during extraction phase: continuous real-time monitoring of CO and MAP was conducted throughout the mechanical heartworm extraction process. Data acquisition was performed at 5 s intervals, and motion artifacts were excluded based on signal quality. This phase spanned from the start of worm retrieval to the completion of extraction; (3) post-extraction stabilization phase: following the completion of heartworm extraction, CO and MAP measurements continued during a 10–15 min stabilization period. Again, 15 measurements were collected per patient to assess cardiovascular recovery and detect any delayed hemodynamic changes. This structured protocol allowed for comprehensive evaluation of hemodynamic trends associated with heartworm removal. Post-extraction stabilization measurements were taken after completion of the heartworm extraction procedure to evaluate recovery and detect any residual hemodynamic instability. The heartworm extraction procedure was performed using a basket retrieval device (S & G Biotech Inc., Seoul, Republic of Korea) under fluoroscopic guidance.

In the GD group, ICG monitoring was also conducted in three phases: pre-decompression, post-decompression, and post-stabilization. Pre-decompression measurements established baseline hemodynamic parameters before the intervention. Post-decompression measurements were recorded immediately after the procedure to assess immediate CO changes. Post-stabilization monitoring continued until hemodynamic parameters stabilized. GD was performed via orogastric intubation, followed by surgical correction when necessary. In one GD case, hemodynamic measurements were performed after initial fluid therapy and successful orogastric decompression, during a period of clinical stabilization. All measurements were completed within a 15 min using a standardized three-phase protocol (pre-decompression, during decompression, and immediate post-decompression). Approximately 30 min after the final measurement, the dog developed recurrent gastric distension and was diagnosed with gastric dilatation-volvulus (GDV), requiring emergency surgical intervention. The procedure included exploratory laparotomy, gastric de-rotation, viability assessment of abdominal organs, and prophylactic gastropexy.

2.3. Statistical Analysis

Statistical analyses were conducted using GraphPad Prism version 10.3.0 (GraphPad Software, San Diego, CA, USA). The primary focus of this study was to evaluate within-group hemodynamic changes over time during therapeutic interventions, rather than to perform direct comparisons between different disease groups or against controls. Normality of the data distributions was assessed using the Shapiro–Wilk test. Since the assumption of normality was satisfied ($p > 0.05$), repeated measures ANOVA (RM-ANOVA) was applied for within-group comparisons among three time points (Pre, Post, and Stabilization).

When the assumption of sphericity was violated, the Geisser–Greenhouse correction was applied. Tukey's post hoc test was used for multiple pairwise comparisons. Multiplicity-adjusted p-values were reported, and statistical significance was considered at $p < 0.05$. Additionally, 95% confidence intervals (CIs) were calculated where appropriate to enhance the interpretation of clinical significance. Non-parametric analyses were conducted using the Friedman test to validate findings, followed by Wilcoxon signed-rank tests for pairwise comparisons. To account for inter-individual variability, changes in cardiac index (CI) were expressed as relative change percentages, calculated using the following formula:

$$Relative\ Change\ (\%) = \frac{Post - treatment\ Value - Pre - treatment\ Value}{Pre - treatment\ Value} \times 100 \quad (1)$$

A p-value of <0.05 was considered statistically significant.

3. Results

3.1. Post-Procedure Outcomes and Recovery

All NC dogs successfully underwent the procedures without complications and recovered uneventfully from anesthesia without any adverse effects. In the HW group, post-procedural echocardiographic examination confirmed a reduction in heartworm burden, allowing the dogs to proceed with the standard heartworm treatment protocol [24]. Representative radiographic and echocardiographic findings for diseased dogs (HW and GD groups) are shown (Figure 1). In the GD group, all dogs were hospitalized for continuous monitoring after decompression, with no recurrence of GD observed. The average hospitalization period was 5.75 ± 0.63 days, and at the three-month follow-up examination, no abnormalities were detected, confirming successful long-term recovery. There were no significant differences in age, weight, or gender distribution between the three groups.

3.2. Hemodynamic Changes Following Intervention

To evaluate the effects of treatment on hemodynamic parameters, cardiac output (CO), cardiac index (CI), and mean arterial pressure (MAP) were measured at baseline, during the procedure, and at the final phase. Hemodynamic data expressed as mean ± SEM with 95% confidence intervals (CIs) are summarized (Table 1). In addition to parametric analyses, non-parametric analyses using the Friedman test and Wilcoxon signed-rank tests were conducted to validate the robustness of the findings. The results confirmed significant improvements in cardiac output (CO) and cardiac index (CI) after intervention, consistent with the parametric analysis outcomes (Table 2). Furthermore, time-course changes in cardiac index (CI), expressed as percentage changes relative to baseline, and their corresponding statistical significance are summarized (Table 3). In the normal control (NC) group, stable anesthesia and recovery were observed throughout the procedure. Hemodynamic parameters, including CO, CI, and MAP, remained stable without significant fluctuations ($p > 0.05$), indicating well-maintained physiological stability under anesthesia and post-procedural recovery. In the heartworm disease (HW) group, a significant improvement in cardiac index (CI) was observed after heartworm removal compared to baseline. Post-procedural echocardiographic examination confirmed a reduction in heartworm burden, allowing the dogs to proceed with the standard heartworm treatment protocol [16]. However, changes in cardiac output (CO) and mean arterial pressure (MAP) did not reach statistical significance. In the gastric dilatation (GD) group, cardiac output (CO) and cardiac index (CI) increased significantly after decompression, indicating substantial improvement in cardiac performance. Although mean arterial pressure (MAP) also tended to increase, the change was not statistically significant. The time-course changes in cardiac index (CI) within each disease group, illustrating significant improvements after intervention, are shown (Figure 3).

Table 1. Summary of patient characteristics.

Group	Parameter	Baseline	Mid-Point	Final-Point
Normal Control ($n = 6$)	CO (L/min)	1.217 ± 0.053 (95% CI: 1.113–1.321)	1.212 ± 0.049 (95% CI: 1.116–1.308)	1.244 ± 0.054 (95% CI: 1.138–1.350)
	CI (L/min/m2)	2.706 ± 0.101 (95% CI: 2.508–2.904)	2.689 ± 0.090 (95% CI: 2.513–2.865)	2.760 ± 0.098 (95% CI: 2.568–2.952)
	MAP (mmHg)	86.689 ± 0.860 (95% CI: 85.003–88.375)	85.478 ± 0.710 (95% CI: 84.086–86.870)	85.967 ± 0.659 (95% CI: 84.675–87.259)
Heartworm Disease (HW) ($n = 6$)	CO (L/min)	1.563 ± 0.529 (95% CI: 0.526–2.600)	1.660 ± 0.527 (95% CI: 0.627–2.693)	1.786 ± 0.581 (95% CI: 0.647–2.925)
	CI (L/min/m2)	2.844 ± 0.088 (95% CI: 2.636–3.052)	2.989 ± 0.080 (95% CI: 2.678–3.114)	3.263 ± 0.086 * (95% CI: 2.652–3.068)
	MAP (mmHg)	76.067 ± 0.861 (95% CI: 74.379–77.755)	76.483 ± 0.851 (95% CI: 74.815–78.151)	76.517 ± 0.685 (95% CI: 75.174–77.860)
Gastric Dilatation (GD) ($n = 4$)	CO (L/min)	1.775 ± 0.108 (95% CI: 1.544–2.006)	2.220 ± 0.132 (95% CI: 1.961–2.479)	2.288 ± 0.131 (95% CI: 2.031–2.545)
	CI (L/min/m2)	2.706 ± 0.137 (95% CI: 2.437–2.975)	3.397 ± 0.148 * (95% CI: 3.107–3.687)	3.500 ± 0.147 * (95% CI: 3.212–3.788)
	MAP (mmHg)	56.567 ± 1.212 (95% CI: 54.191–58.943)	69.700 ± 0.790 (95% CI: 68.152–71.248)	70.967 ± 0.686 (95% CI: 69.622–72.312)

Values are presented as mean ± SEM (standard error of mean), with 95% confidence intervals (CIs) provided separately below each value. Time points: normal control group—stabilization, process, end phase; heartworm disease (HW) group—pre-extraction, during extraction, post-extraction; gastric dilatation (GD) group—pre-decompression, post-decompression, stabilization. * $p < 0.05$ compared to baseline within the same group. CO = cardiac output; CI = cardiac index; MAP = mean arterial pressure.

Table 2. Non-parametric analysis results (Friedman test and Wilcoxon signed-rank test).

Group	Parameter	Friedman Test p-Value	Significant Pairwise Comparisons (Wilcoxon Test)	Notes
Heartworm Disease (HW) ($n = 6$)	CO (L/min)	$p < 0.05$	Pre vs. During ($p < 0.05$), Pre vs. Post ($p < 0.05$), During vs. Post (n.s.)	Increase from Pre to Post
	CI (L/min/m2)	$p < 0.05$	Pre vs. During ($p < 0.05$), Pre vs. Post ($p < 0.05$), During vs. Post (n.s.)	Increase from Pre to Post
	MAP (mmHg)	0.84	Not applicable	No significant difference
Gastric Dilatation (GD) ($n = 4$)	CO (L/min)	$p < 0.01$	Pre vs. Post ($p < 0.01$), Pre vs. Stabilization ($p < 0.01$), Post vs. Stabilization (n.s.)	Increase after decompression
	CI (L/min/m2)	$p < 0.01$	Pre vs. Post ($p < 0.01$), Pre vs. Stabilization ($p < 0.01$), Post vs. Stabilization (n.s.)	Increase after decompression
	MAP (mmHg)	$p < 0.01$	Pre vs. Post ($p < 0.01$), Pre vs. Stabilization ($p < 0.01$), Post vs. Stabilization (n.s.)	Increase after decompression

Summary of the non-parametric analysis results using the Friedman test and post hoc Wilcoxon signed-rank tests. Significant differences between specific time points are indicated. Abbreviations: HW: heartworm disease group; GD: gastric dilatation group; CO: cardiac output; CI: cardiac index; MAP: mean arterial pressure; n.s.: not significant.

Table 3. Time-course changes in cardiac index and statistical significance.

Group	Baseline CI	Mid-Point Change (%)	*p*-Value (Mid-Point)	Final-Point Change (%)	*p*-Value (Final-Point)
Normal Control (n = 6)	2.706 ± 0.101	0.40	>0.05	2.12	>0.05
Heartworm Disease (n = 6)	2.844 ± 0.088	5.09	>0.05	14.71	**0.0102** *
Gastric Dilatation (n = 4)	2.706 ± 0.137	28.26	**0.0184** *	33.81	**0.0336** *

CI values are presented as mean ± SEM (L/min/m^2). Time points: normal control—stabilization→process→end phase; HW group—pre-extraction→during extraction→post-extraction; GD group—pre-decompression→post-decompression→stabilization. Relative change (%) = (post-treatment value—pre-treatment value)/pre-treatment value × 100. (*) Bold values indicate statistically significant changes ($p < 0.05$).

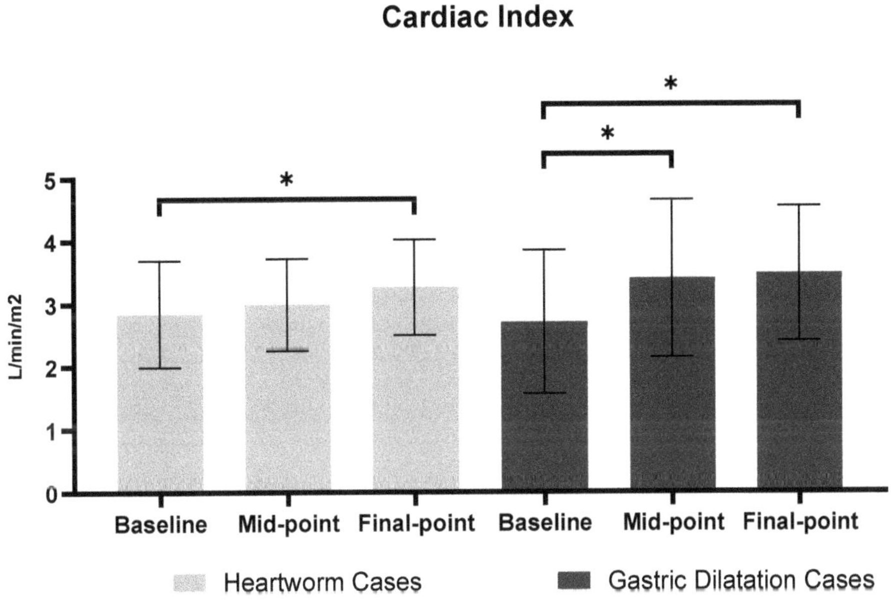

Figure 3. Cardiac Index comparison between heartworm and gastric dilatation cases across three time points. Values are presented as mean ± SEM (standard error of mean). For Heartworm cases, time points represent: baseline (anesthesia stabilization), Mid-point (during extraction), and Final point (after extraction). For gastric dilatation cases, time points represent: baseline (pre-decompression), mid-point (post-decompression), and final-point (stabilization). Asterisks (*) indicate statistically significant differences between time points ($p < 0.05$).

4. Discussion

This study evaluated the feasibility of non-invasive real-time CO monitoring in dogs with HW and GD using ICG. By establishing the baseline CO and MAP values from healthy dogs, this study enabled a comparative assessment of hemodynamic changes before and after therapeutic interventions in HW and GD cases. The incorporation of a normal control (NC) group in this study design served dual methodological purposes. Primarily, it established reference ranges for ICG-derived hemodynamic parameters in clinically healthy canines under standardized conditions. These normative data are particularly valuable given the paucity of established reference intervals for ICG measurements in veterinary medicine. The observed stability in hemodynamic parameters among control subjects (CI: 2.706 ± 0.101 L/min/m^2 at baseline with nonsignificant variations of 0.40% and 2.12%

at subsequent time points) provides critical context for interpreting pathophysiological alterations in disease states.

Our methodological approach prioritized within-group hemodynamic changes in response to therapeutic interventions. The comparable baseline CI values observed across the three groups merit careful consideration from multiple perspectives. In chronic heartworm disease, physiologic compensation—such as increased heart rate and myocardial contractility—may maintain near-normal resting cardiac output despite underlying pulmonary vascular compromise. Similarly, during the early stages of gastric dilatation, cardiovascular dysfunction may be present without substantial reductions in CI at the time of measurement, potentially due to initial resuscitative measures administered prior to ICG monitoring. Additionally, the intrinsic characteristics of impedance cardiography warrant consideration regarding the potential overestimation of CI in disease groups. Anatomical alterations in heartworm disease or gastric dilatation (e.g., increased pulmonary vascular resistance and elevated intra-abdominal pressure) might affect impedance measurements, yielding CI values that appear higher than the actual hemodynamic status. These considerations further support the utility of ICG as a trend-monitoring tool rather than a method for absolute hemodynamic quantification.

Furthermore, the consistent measurements obtained across multiple time points in the NC group (stabilization, sedation, and anesthesia stabilization phases) demonstrate the reliability and reproducibility of ICG technology for serial hemodynamic assessments, which is commonly referred to as "trending ability" in hemodynamic monitoring literature [4,12,25]. This technical validation aspect is essential when evaluating a monitoring modality intended for continuous assessment during dynamic clinical scenarios.

The hemodynamic stability observed in the NC group stands in marked contrast to the significant temporal variations detected in both intervention groups. The quantifiable improvements in CI following heartworm extraction (+14.71%, p = 0.0102) and gastric decompression (+33.81%, p = 0.0336) represent statistically significant hemodynamic responses to therapeutic interventions rather than measurement artifacts. This differentiation between pathophysiological stability and intervention-induced change substantiates the clinical utility of ICG for real-time assessment of cardiovascular responses to treatment.

These findings suggest that ICG not only provides absolute hemodynamic values but also reliably tracks directional changes in cardiac performance over time. This trending capability may prove particularly valuable in emergency and critical care settings, where rapid the assessment of treatment efficacy and early detection of hemodynamic deterioration can significantly impact patient outcomes.

In both heartworm extraction and gastric decompression procedures, systemic blood pressure alone may be insufficient to determine the optimal endpoint of intervention. In our study, several HW cases exhibited a measurable increase in cardiac output following parasite removal yet mean arterial pressure (MAP) remained suboptimal or fluctuated—a finding consistent with prior experimental studies involving heartworm-associated hemodynamic changes [17,26]. Similarly, in GD cases, a significant improvement in cardiac index (28.53%, p = 0.0336) following decompression was not proportionally reflected in MAP values. This dissociation underscores the limitations of relying solely on arterial pressure for clinical decision-making. The use of impedance cardiography (ICG) allowed continuous tracking of cardiac performance trends throughout both procedures, providing critical real-time data beyond conventional monitoring parameters. These additional insights enabled more comprehensive evaluation of the physiological response to treatment and supported clinicians in determining appropriate endpoints for intervention. Our findings suggest that ICG may serve as a valuable adjunctive monitoring tool in emergency and critical care settings, especially for conditions such as heartworm disease and gastric dilatation, where traditional vital signs may inadequately reflect the true hemodynamic status.

The PhysioFlow® system addresses several technical challenges inherent to impedance cardiography. Importantly, the system's HD-Z™ filtering technology mitigates the potential interference between respiratory and cardiac cycles by analyzing impedance waveform morphology rather than relying on baseline thoracic impedance (Z0). The system samples at 250 Hz, capturing multiple cardiac cycles within each 5 s measurement interval, which enables reliable trend monitoring despite varying respiratory patterns during anesthesia. This signal processing approach effectively isolates cardiac-related impedance changes from respiratory artifacts, maintaining measurement consistency even during mechanical ventilation. While factors such as electrode placement, thoracic conformation, and tissue characteristics may influence absolute values, the system's focus on waveform analysis rather than absolute impedance values helps maintain reliable trend tracking. This capability was evident in our study, where consistent hemodynamic response patterns were observed across different clinical scenarios, supporting ICG's utility as a trend-monitoring tool during critical interventions.

Non-invasive methods for measuring cardiac output (CO) are becoming increasingly relevant, particularly in small animal research where minimizing procedural risks is a priority [1,25,27,28]. Among these, impedance cardiography (ICG) has gained attention for its ability to track stroke volume (SV) and CO in real-time by detecting thoracic electrical conductivity changes throughout the cardiac cycle [29]. This technique has proven valuable in human neonatal and pediatric patients, where continuous, non-invasive monitoring is essential [30]. While its clinical application in veterinary medicine is still under investigation, ongoing studies suggest its potential for accurate and practical CO assessment [4,9,11,16,31–34].

In heartworm (HW) disease, CO measurement using pulmonary artery catheterization is often impractical, as the invasive nature of the procedure can interfere with parasite extraction. Additionally, HW-related cardiovascular alterations, such as pulmonary hypertension and right ventricular dysfunction, necessitate continuous CO monitoring to evaluate hemodynamic responses. Similarly, gastric dilatation (GD) presents unique challenges, as decompression induces sudden shifts in venous return and CO. In GD cases, rapid intervention is critical, making the use of invasive monitoring devices impractical due to the time required for insertion and stabilization. In contrast, ICG can be set up within 10 min, with minimal disruption to ongoing procedures [4]. Furthermore, its non-invasive nature ensures minimal interference with other critical interventions, allowing for seamless integration into emergency treatment protocols. Therefore, in this study, ICG was selected as a non-invasive tool capable of providing real-time CO monitoring in both HW and GD cases, allowing for continuous assessment of cardiovascular stability during critical interventions.

In the GD group, a moderate CO decrease was observed before decompression, likely due to impaired venous return caused by increased intra-abdominal pressure. Following decompression, ICG detected a rapid CO recovery, supporting the effectiveness of decompression in restoring hemodynamic stability. This immediate improvement in CO can be attributed not only to the relief of intra-abdominal pressure but also to the concurrent administration of intravenous fluid therapy. The increased venous return from aggressive fluid resuscitation likely contributed to the rapid hemodynamic restoration observed post-decompression [35,36].

Previous studies have demonstrated ICG's potential for non-invasive CO monitoring in human and veterinary medicine [13], yet its application in HW and GD cases remains underexplored. Traditional CO measurement methods, such as Doppler echocardiography and thermodilution, provide accurate assessments but are often invasive, operator-dependent, or impractical for continuous monitoring [8,15]. Unlike these techniques, ICG offers real-time CO measurement with minimal procedural burden, making it more suitable for acute and emergency settings [11].

In this study, the ICG was able to detect transient CO fluctuations during HW extraction, a feature that is typically undetected by static measurement techniques. Our finding suggests that ICG may be used to complement existing hemodynamic monitoring tools by providing continuous real-time data, particularly in conditions with dynamic cardiovascular changes.

This study has limitations that should be addressed in future research. The small sample size (GD: 4, HW: 6) limits the generalizability of the findings. Future studies with larger and more diverse populations are essential to validate the broader applicability of ICG. Additionally, reliable data acquisition from ICG requires skillful electrode placement and stable signal quality. Improper placement or signal interference during emergencies may lead to inaccurate readings, which could adversely affect clinical outcomes. Further advancements in electrode design and signal-processing algorithms are necessary to enhance data stability and accuracy.

Additionally, ICG's reliance on thoracic impedance and ECG-based calculations may pose challenges in patients with arrhythmias, such as premature ventricular contractions (VPCs), potentially affecting data accuracy [22,37]. To improve the clinical applicability of ICG, the development of advanced algorithms to provide reliable measurements under arrhythmic conditions is worth considering.

Importantly, while ICG provides valuable hemodynamic data, it should be viewed as an adjunctive tool only. Clinical expertise is indispensable in interpreting ICG data in the context of a patient's overall condition, history, and other monitoring parameters. While ICG is a powerful adjunct to guide therapeutic interventions, effective decision-making requires integrating all available clinical information, supported by a skilled medical team [37,38]. In this study, transient arrhythmic events were observed in some GD cases, particularly during the initial stabilization phase. However, these episodes were short-lived, and stable sinus rhythm was restored prior to ICG measurement. All reported hemodynamic data were obtained during periods of rhythm stability, minimizing the influence of arrhythmias on the interpretation of CI and MAP trends.

Future research should investigate the efficacy of ICG in larger and more diverse populations across various clinical conditions. By addressing these challenges, ICG has the potential to become a cornerstone technology for real-time hemodynamic monitoring, particularly for dynamic cardiovascular conditions.

5. Conclusions

This study explored the potential utility of non-invasive impedance cardiography (ICG) for real-time CO monitoring in canine HW and GD cases. By providing immediate trend-based hemodynamic data, ICG demonstrated promise as a monitoring tool that could support clinical decision-making during interventions characterized by acute CO fluctuations. The consistent detection of hemodynamic improvement following therapeutic interventions suggests ICG may offer valuable insights into cardiovascular responses, though larger studies are needed to confirm these preliminary findings.

ICG's non-invasive application, rapid implementation, and concurrent monitoring of CO alongside standard MAP measurements provides more comprehensive hemodynamic assessment than conventional monitoring alone. This combined monitoring approach shows potential clinical utility in emergencies and critical care settings where complete cardiovascular data can inform intervention strategies. This pilot study suggests that integrating ICG with traditional monitoring parameters may offer clinicians additional insights into cardiovascular status during critical interventions in canine patients. With further validation through expanded clinical trials, this combined hemodynamic assessment approach could enhance monitoring capabilities in veterinary emergencies and routine care.

Author Contributions: D.K. and S.-G.P. wrote the manuscript and collected the residual samples and data; Conceptualization, D.K.; writing—original draft preparation, D.K.; project administration, M.S.K.; writing—review and editing, M.S.K. All authors have read and agreed to the published version of the manuscript.

Funding: This study was supported by the National Research Foundation of Korea (NRF) grant funded by the Korean Government (MSIT) (No. 2022R1F1A1065215).

Institutional Review Board Statement: The Institutional Animal Care and Use Committee (IACUC) of Seoul National University SNU-200226-7-2, SNU-240115-9-1.

Informed Consent Statement: The cases were client-owned, and the clients gave their consent to the publication of information about the patients in the case report.

Data Availability Statement: The data presented in this study are available in this manuscript.

Acknowledgments: The authors extend their gratitude to mdksystem Korea for providing the rental of PhysioFlow†TM and to the staff responsible for the care of dogs during treatment and hospitalization. We sincerely thank Joungsun Moon and I-se O for their invaluable support and contributions to this study.

Conflicts of Interest: The authors declare no conflict of interest.

References

1. Nguyen, L.S.; Squara, P. Non-Invasive Monitoring of Cardiac Output in Critical Care Medicine. *Front. Med.* **2017**, *4*, 200. [CrossRef] [PubMed]
2. Argyroudis, E. Principles of Hemodynamic Monitoring. *Rev. Clin. Pharmacol. Pharmacokinet. Int. Ed.* **2001**, *15*, 207–210.
3. Platt, M.J.; Huber, J.S.; Brunt, K.R.; Simpson, J.A. Pulmonary Flow as an Improved Method for Determining Cardiac Output in Mice after Myocardial Infarction. *J. Am. Soc. Echocardiogr.* **2017**, *30*, 612–623.e1. [CrossRef]
4. Paranjape, V.V.; Garcia-Pereira, F.L.; Menciotti, G.; Saksena, S.; Henao-Guerrero, N.; Ricco-Pereira, C.H. Evaluation of Electrical Cardiometry for Measuring Cardiac Output and Derived Hemodynamic Variables in Comparison with Lithium Dilution in Anesthetized Dogs. *Animals* **2023**, *13*, 2362. [CrossRef]
5. Cabrales, P.; Acero, C.; Intaglietta, M.; Tsai, A.G. Measurement of the Cardiac Output in Small Animals by Thermodilution. *Microvasc. Res.* **2003**, *66*, 77–82. [CrossRef]
6. Paranjape, V.V.; Pereira, F.L.G.; Menciotti, G.; Saksena, S.; Guerrero, N.H.; Pereira, C.H.R. Agreement of Cardiac Output Measurements by Esophageal Doppler and Transesophageal Echocardiography with Intermittent Pulmonary Artery Thermodilution during Pharmacologic Manipulation of Hemodynamics in Anesthetized Dogs. *Am. J. Vet. Res.* **2023**, *84*, 1–11. [CrossRef]
7. Paranjape, V.V.; Henao-Guerrero, N.; Menciotti, G.; Saksena, S. Performance of Four Cardiac Output Monitoring Techniques vs. Intermittent Pulmonary Artery Thermodilution during a Modified Passive Leg Raise Maneuver in Isoflurane-Anesthetized Dogs. *Front. Vet. Sci.* **2023**, *10*, 1238549. [CrossRef]
8. Sodolski, T.; Kutarski, A. Impedance Cardiography: A Valuable Method of Evaluating Haemodynamic Parameters. *Folia Cardiol.* **2007**, *14*, 115–126.
9. Mantovani, M.M.; Fantoni, D.T.; Gimenes, A.M.; de Castro, J.R.; Flor, P.B.; Ida, K.K.; Schwartz, D.S. Clinical Monitoring of Cardiac Output Assessed by Transoesophageal Echocardiography in Anaesthetised Dogs: A Comparison with the Thermodilution Technique. *BMC Vet. Res.* **2017**, *13*, 325. [CrossRef]
10. Li, H.; Abaei, A.; Metze, P.; Just, S.; Lu, Q.; Rasche, V. Technical Aspects of in Vivo Small Animal CMR Imaging. *Front. Phys.* **2020**, *8*, 1–26. [CrossRef]
11. Payseur, J.D.; Rigney, J.J.; Turner, S.L.; Wu, X.J.; Murphy, D.J.; Rossman, E.I. Evaluation of a Method Utilizing PhysioFlow® a Novel Signal Morphology-Based Form of Impedance Cardiography, to Measure Cardiac Output in the Conscious Beagle. *J. Pharmacol. Toxicol. Methods* **2016**, *81*, 115–119. [CrossRef] [PubMed]

12. Mutoh, T.; Sasaki, K.; Yamamoto, S.; Yasui, N.; Ishikawa, T.; Taki, Y. Performance of Electrical Velocimetry for Noninvasive Cardiac Output Measurements in Perioperative Patients after Subarachnoid Hemorrhage. *J. Neurosurg. Anesthesiol.* **2019**, *31*, 422–427. [CrossRef] [PubMed]
13. Paranjape, V.V.; Henao-Guerrero, N.; Menciotti, G.; Saksena, S.; Agostinho, M. Agreement between Electrical Cardiometry and Pulmonary Artery Thermodilution for Measuring Cardiac Output in Isoflurane-Anesthetized Dogs. *Animals* **2023**, *13*, 1420. [CrossRef] [PubMed]
14. Sasaki, K.; Mutoh, T.T.; Mutoh, T.T.; Kawashima, R.; Tsubone, H. Electrical Velocimetry for Noninvasive Cardiac Output and Stroke Volume Variation Measurements in Dogs Undergoing Cardiovascular Surgery. *Vet. Anaesth. Analg.* **2017**, *44*, 7–16. [CrossRef]
15. Schmidt, C.; Theilmeier, G.; Van Aken, H.; Korsmeier, P.; Wirtz, S.P.; Berendes, E.; Hoffmeier, A.; Meissner, A. Comparison of Electrical Velocimetry and Transoesophageal Doppler Echocardiography for Measuring Stroke Volume and Cardiac Output. *Br. J. Anaesth.* **2005**, *95*, 603–610. [CrossRef]
16. Peng, Z.; Critchley, L.A.H.; Fok, B.S.P.; James, A.E. Evaluation of Impedance Based Indices of Cardiac Contractility in Dogs. *J. Clin. Monit. Comput.* **2004**, *18*, 103–109. [CrossRef]
17. Kim, D.; Park, S.G.; Kim, M. Interventional Heartworm Extraction in Two Dogs: The Clinical Application of Impedance Cardiography. *Animals* **2023**, *13*, 3127. [CrossRef]
18. Quandt, J. Anesthesia for the Dog with Heartworm Disease: A Brief, Practical Review. *Parasites Vectors* **2023**, *16*, 151. [CrossRef]
19. Lee, Y.; Kim, H.; Yoon, K.; Park, J.; Oh, Y.; Oh, M.; Ban, J.; Lee, M.; Kim, H.; Kang, B.T.; et al. Clinical Application of a Newly Developed Basket Device for Interventional Heartworm Extraction in Three Dogs. *J. Vet. Clin.* **2022**, *39*, 177–184. [CrossRef]
20. Homer, A. Cardiovascular Monitoring of the Canine Gastric Dilatation Volvulus Patient. *Vet. Nurse* **2020**, *11*, 24–28. [CrossRef]
21. Horne, W.A.; Gilmore, D.R.; Dietze, A.E.; Freden, G.O.; Short, C.E. Effects of Gastric Distention-Volvulus on Coronary Blood Flow and Myocardial Oxygen Consumption in the Dog. *Am. J. Vet. Res.* **1985**, *46*, 98–104. [CrossRef] [PubMed]
22. Sharp, C.R.; Rozanski, E.A. Cardiovascular and Systemic Effects of Gastric Dilatation and Volvulus in Dogs. *Top. Companion Anim. Med.* **2014**, *29*, 67–70. [CrossRef] [PubMed]
23. MacPhail, C.M.; Monnet, E.; Pelsue, D.H.; Gaynor, J.S. Evaluation of Cardiac Performance of the Dog after Induction of Portal Hypertension and Gastric Ischemia. *J. Vet. Emerg. Crit. Care* **2006**, *16*, 192–198. [CrossRef]
24. Romito, G.; Pane, E.; Guglielmini, C.; Poser, H.; Valente, C.; Paradies, P.; Castagna, P.; Mazzoldi, C.; Cipone, M. Efficacy and Tolerability of the American Heartworm Society Therapeutic Protocol in Dogs Affected by Heartworm Disease without Caval Syndrome. *J. Small Anim. Pract.* **2024**, *65*, 39–46. [CrossRef]
25. Briganti, A.; Evangelista, F.; Centonze, P.; Rizzo, A.; Bentivegna, F.; Crovace, A.; Staffieri, F. A Preliminary Study Evaluating Cardiac Output Measurement Using Pressure Recording Analytical Method (PRAM) in Anaesthetized Dogs. *BMC Vet. Res.* **2018**, *14*, 72. [CrossRef]
26. Kitoh, K.; Mikami, C.; Kitagawa, H.; Sasaki, Y. Hemodynamic Alterations in Dogs with Shock Induced by Intravenous Injection of Heartworm Extract. *J. Vet. Med. Sci.* **2001**, *63*, 179–182. [CrossRef]
27. Lopes, P.C.F.; Sousa, M.G.; Camacho, A.A.; Carareto, R.; Nishimori, C.T.D.; Santos, P.S.P.; Nunes, N. Comparison between Two Methods for Cardiac Output Measurement in Propofol-anesthetized Dogs: Thermodilution and Doppler. *Vet. Anaesth. Analg.* **2010**, *37*, 401–408. [CrossRef]
28. Haryadi, D.G.; Orr, J.A.; Kuck, K.; McJames, S.; Westenskow, D.R. Partial CO_2 Rebreathing Indirect Fick Technique for Non-Invasive Measurement of Cardiac Output. *J. Clin. Monit. Comput.* **2000**, *16*, 361–374. [CrossRef]
29. Bernstein, D.P.; Lemmens, H.J.M. Stroke Volume Equation for Impedance Cardiography. *Med. Biol. Eng. Comput.* **2005**, *43*, 443–450. [CrossRef]
30. Grollmuss, O.; Gonzalez, P. Non-Invasive Cardiac Output Measurement in Low and Very Low Birth Weight Infants: A Method Comparison. *Front. Pediatr.* **2014**, *2*, 16. [CrossRef]
31. Sasaki, K.; Mutoh, T.; Yamamoto, S.; Taki, Y.; Kawashima, R. Comparison of Noninvasive Dynamic Indices of Fluid Responsiveness among Different Ventilation Modes in Dogs Recovering from Experimental Cardiac Surgery. *Med. Sci. Monit.* **2018**, *24*, 7736–7741. [CrossRef] [PubMed]
32. Sasaki, K.; Mutoh, T.; Mutoh, T.; Taki, Y.; Kawashima, R. Noninvasive Stroke Volume Variation Using Electrical Velocimetry for Predicting Fluid Responsiveness in Dogs Undergoing Cardiac Surgery. *Vet. Anaesth. Analg.* **2017**, *44*, 719–726. [CrossRef] [PubMed]
33. Sasaki, K.; Yamamoto, S.; Mutoh, T. Noninvasive Assessment of Fluid Responsiveness for Emergency Abdominal Surgery in Dogs with Pulmonary Hypertension: Insights into High-Risk Companion Animal Anesthesia. *PLoS ONE* **2020**, *15*, e0241234. [CrossRef]
34. Tremper, K.K.; Hufstedler, S.M.; Barker, S.J.; Zaccari, J.; Harris, D.; Anderson, S.; Roohk, V. Continuous Noninvasive Estimation of Cardiac Output by Electrical Bioimpedance: An Experimental Study in Dogs. *Crit. Care Med.* **1986**, *14*, 231–233. [CrossRef]

35. Mazzaferro, E.M.; Monnet, E. Gastric Dilatation Volvulus. In *Small Animal Soft Tissue Surgery*; Wiley: Hoboken, NJ, USA, 2023; pp. 53–74. [CrossRef]
36. Monnet, E. Gastric Dilatation-Volvulus Syndrome in Dogs. *Vet. Clin. N. Am. Small Anim. Pract.* **2003**, *33*, 987–1005. [CrossRef]
37. Bour, J.; Kellett, J. Impedance Cardiography—A Rapid and Cost-Effective Screening Tool for Cardiac Disease. *Eur. J. Intern. Med.* **2008**, *19*, 399–405. [CrossRef]
38. Van De Water, J.M.; Miller, T.W.; Vogel, R.L.; Mount, B.E.; Dalton, M.L. Impedance Cardiography the next Vital Sign Technology? *Chest* **2003**, *123*, 2028–2033. [CrossRef]

veterinary sciences

MDPI

Article

Kinesiology Taping Does Not Affect Tarsal Joint Motion During Selected Exercises in Dogs †

Rebecca C. Noel [1], Leann M. Shaw [2], Nicholas H. Millis [3], Krysta Janas [4] and Darryl L. Millis [5,*]

1 Metairie Small Animal Hospital, New Orleans, LA 70005, USA; bnoeldvm@gmail.com
2 VCA California Veterinary Specialists, Murrieta, CA 92563, USA; leannshawdvm@gmail.com
3 Veterinary Care and Specialty Group, Chattanooga, TN 37408, USA; nmillisdvm@gmail.com
4 BluePearl Pet Hospital Gwinnett, Lawrenceville, GA 30043, USA; krysta.janas@gmail.com
5 Department of Small Animal Clinical Sciences, College of Veterinary Medicine, University of Tennessee, Knoxville, TN 37919, USA
* Correspondence: dmillis@utk.edu; Tel.: +1-865-974-8387

† This paper is an extended version of the abstract Effect of kinesiology taping on gait and selected exercises in dogs published in the Proceedings of the 11th International Association of Veterinary Rehabilitation and Physical Therapy. Cambridge, England, Acta Veterinaria Scandinavica 2023, 65(Suppl 1):O.03 https://doi.org/10.1186/s13028-023-00706-w.

Simple Summary: Kinesiology taping has increased in use in human and veterinary medicine, both for athletic performance and treatment of various musculoskeletal and neurologic conditions. Studies in people suggest a mild effect for many uses. There are very few studies regarding kinesiology taping in veterinary medicine. Prospective evaluation of its effect on gait and mobility is essential to evaluate the efficacy of kinesiology taping. The objective of this study was to evaluate the effect of kinesiology taping applied to the tarsal joint and its effect on selected exercises in dogs using kinetic and kinematic gait analysis. In normal dogs, kinesiology taping had no effect on weightbearing forces at a walk and trot, and motion of the stifle and tarsal joints while walking, trotting, or stepping over cavaletti rails. Kinesiology taping should be assessed in other joints and in dogs with neurologic or orthopedic conditions.

Abstract: Background: Use of kinesiology taping has increased in veterinary medicine, yet there are few studies of its effects. The purpose of this study was to evaluate the effect of kinesiology tape placed on the dorsal surface of the tarsus on kinetic and kinematic gait characteristics while performing selected exercises in dogs using 3D motion capture and force platform analysis of gait. Methods: Ten clinically normal, healthy, adult mixed-breed dogs were recruited for this study. Reflective markers were applied to the skin of the left and right pelvic limbs of each dog. Eight infrared cameras were positioned around a 13 m platform containing a force platform. Dogs were walked, trotted, and led over cavaletti rails with and without kinesiology tape extending from the distal tibia, crossing the cranial surface of the tarsus, and extending to the dorsal surface of the metatarsals. The evaluations were repeated 2 h later. Maximum flexion and extension in the sagittal plane and maximum angular acceleration and velocity were measured at the tarsus and stifle for each dog. Peak vertical (Z_{Peak}), braking (Y_A), and propulsion (Y_B) forces were determined as a percentage of body weight while walking and trotting. Results: Ground-reaction forces were greater at the trot compared to the walk, and stifle and tarsal flexion were greater with stepping over cavaletti rails as compared to walking and trotting. There were no differences in kinematic or kinetic measurements related to the kinesiology tape application, time, or exercise being performed. Conclusions: Kinesiology tape had no effect on tarsal joint gait characteristics while walking, trotting, or stepping over cavaletti rails when applied to the dorsal tarsus of normal dogs.

Keywords: gait analysis; kinematic gait evaluation; kinesiology tape; kinetic gait evaluation

1. Introduction

Kinesiology taping has increased in popularity both in human and veterinary applications since its inception in the 1970s. Today, it is commonly seen in athletes of numerous species in various activities. Application of kinesiology tape to the skin reportedly targets sensory receptors of the cutaneous skin, lifts tissues to provide lymphatic drainage, and improves circulation with simultaneous analgesia via nociceptive transmission inhibition [1]. There may be a benefit in using kinesiology tape for pain management [2–5], although the effects seem to be mild in musculoskeletal conditions, including osteoarthritis pain [6–8]. Additionally, kinesiology tape is postulated to stimulate pathways related to the mechanoreceptive and proprioceptive pathways, thereby modifying joint kinematics [9]. In humans, it is also used to improve athletic performance despite a lack of literary support [10]. However, most studies indicate negligible benefit for muscle strengthening in healthy adult humans [11,12], and there may be few benefits regarding athletic performance in healthy individuals [10]. Some studies have suggested an improvement in muscle fatigue [13–15], while others have shown no improvement in muscle strength or athletic performance [3,10,16,17]. Further, kinesiology taping may have limited effects to reduce swelling after an acute ankle sprain [18,19].

In veterinary medicine, kinesiology tape is used in horses to treat muscular conditions, fascial restrictions, and postural imbalance, although there is little evidence of beneficial effects [9,20–22]. Ramon et al. demonstrated decreased peak vertical force and limited flexion of the fetlock in horses during the swing phase of gait with kinesiology tape applied [22]. A study of the effects of kinesiology tape applied to the forelimbs of horses showed no effect on forelimb kinematics or muscle EMG activity [23]. However, application to the abdominal muscles increased craniocaudal activity of the body [24].

There is very little evidence regarding kinesiology tape application techniques or how it affects joint motion and weight bearing during simple exercises in normal dogs. The authors have used kinesiology tape on several dogs with various pathologies and perceived efficacy in terms of increased joint motion. However, the magnitude of effect and the duration of action have not been evaluated under controlled conditions with objective outcome measures in normal dogs. With the advent of veterinary-specific kinesiology tape, manufacturers claim the tape can last for 5 days and can be used to treat conditions ranging from muscle injury, inflammation, tendonitis, desmitis, and increased joint range of motion (RockTape, https://rocktape.com.au/canine/ (accessed on 3 March 2025)).

The purpose of this study was to investigate gait changes following the application of kinesiology tape on the dorsal canine tarsus. Additionally, tape longevity and application protocol were evaluated. We hypothesized that there would be increased range of motion and decreased peak vertical force between the taped limbs of each dog at the walk, trot, and while walking over cavaletti rails.

2. Materials and Methods

Participants: The study protocol was approved by the Institutional Animal Care and Use Committee at the University of Tennessee, and written informed consent was obtained from owners. Ten client-owned dogs were included in the study reported here. Weight, age, and breed were recorded. An initial baseline evaluation was performed, including orthopedic, neurologic, and general physical examinations. Individuals were excluded for any history of significant mobility disorder, obvious visual lameness, greater than 10%

difference in peak vertical force between forelimbs or between hind limbs at a walk and a trot, if gross joint instability was present, or if there were any other clinically abnormal findings on physical examination. Dogs were between 2 and 10 years of age and weighed between 15 and 50 kg.

Kinematic Data Collection: A 3-dimensional testing space measuring 3 × 1 m was established on a 13 m walkway centered over the force platform. A right-handed orthogonal global coordinate system was established within the 3D testing space with 0,0,0 (X, Y, Z) located on the force platform. On each data-collection day, 8 high-speed data-capture cameras (250 Hz Vero cameras, Vicon Motion Systems Inc., Centennila, CO, USA) were calibrated to the testing space using an active calibration wand (Vicon Motion Systems Inc., Centennila, CO, USA). Markers were tracked, and motion capture data were recorded using commercial software (Vicon Nexus version 2.11, Vicon Motion Systems, Inc.).

Motion capture began with a static calibration with 10–15 frames of data captured, with 23 reflective, spherical markers on each pelvic limb (46 total markers) placed on anatomic landmarks or as rigid femur cluster markers on femurs and tibias as described by Fu et al. (Figure 1) [25]. We used multiple rigid cluster design methods based on experience from animal and human biomechanic studies. A 2 × 2 cm and a 3 × 3 cm 3D-printed arched boards using a 20% infill were constructed for application of cluster markers (Figure 2). The boards were covered with Velcro, and the markers were affixed. The arched boards and reflective markers were then affixed to animals using double-sided medical tape and GLUture (Zoetis Inc., Kalamazoo, MI 49007, USA), if necessary. Hair in the area was clipped if more than 0.5 cm in length.

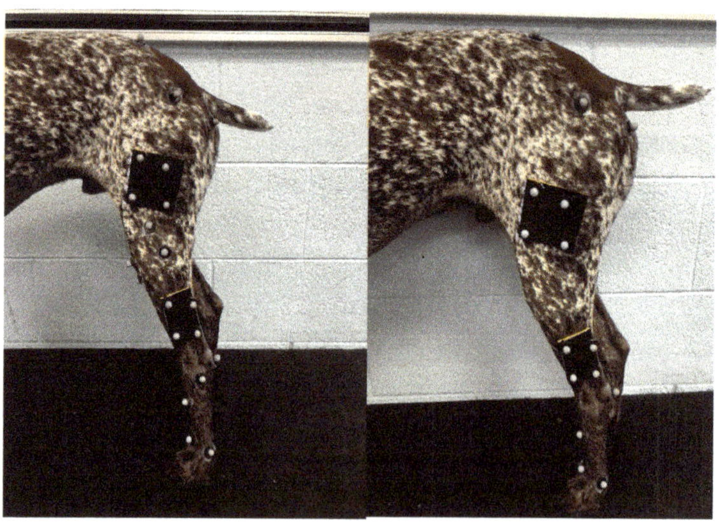

Figure 1. Anatomic (calibration) marker set (**left**) and tracking marker set (**right**).

Following the calibration trial, seven markers over anatomic landmarks were removed for the remaining motion trials (Table 1). These markers were virtually reconstructed from the initial static trial using Vicon Procalc software, version 1.5 (Vicon, Colorado) [25,26]. Each dog was walked, trotted, and walked over cavaletti rails four times with the cluster markers in place. The order of data collection was randomized among the three exercises. Data were acquired from the right and left sides separately.

Figure 2. Rigid cluster design using a 3D-printed curved board. Each cluster was arranged in a rhomboid shape with an approximate distance of 2.5 cm separating each marker from the center of the sphere. Markers were arranged in a non-colinear fashion.

Table 1. List of anatomic markers, ** indicates markers that were removed after kinematic calibration. The greater trochanter was shared between the pelvis and femur segments for calibration but was only used for the pelvis during data acquisition (tracking).

Pelvic Limb Segments	Marker Location (Right and Left Limb)
Pelvis	Iliac wing
	Ischium
Femur	Greater trochanter
	Lateral epicondyle **
	Medial epicondyle **
Tibia	Fibular head **
	Proximal tibial crest **
	Distal tibial crest **
	Lateral malleolus **
	Medial malleolus **
Foot	Point of calcaneus
	Metatarsophalangeal joint 2
	Metatarsophalangeal joint 5
	Proximal tarsometatarsal joint
	Distal tarsometatarsal joint
Cluster Markers (4 each)	Femoral cluster
	Tibial cluster

After these trials were captured, kinesiology tape was applied to the dorsal surface of a randomly selected tarsus from the distal tibial tuberosity to the distal metatarsus. The tape was applied by stretching the tape to 25% of the unstretched tape length (to apply under tension) and placed from proximal to distal (Figure 3). We found it necessary in initial attempts to apply a 3 cm-wide anchoring piece (not stretched) at the proximal and distal ends to help hold the tape in place, according to the manufacturer's instructions (RockTape, Morley, WA, USA). After application, the tape was rubbed for 3 min to activate the adhesive prior to the acclimation period. Dogs were allowed to acclimate to the tape during 5 min of free exercise. Dogs were then walked, trotted, and walked over cavaletti rails to collect kinetic and kinematic data as described, and data were collected 5 min and 2 h after tape application. Cavaletti height was based on the size of the dog, with the cavaletti rail height placed at the junction of the lower and middle third of the antebrachium.

Joint coordinate system: The local coordinate systems (LCS) specific to each segment were designed similarly to Fu et al. [25], with the exception of the foot. The LCS of our foot originated at the caudal aspect of the calcaneus (CALC), with the unit vector of the z-axis defined by the vector between the second and fifth distal metatarsals, the x-axis unit vector was defined by the vector from the caudal calcaneus travelling distally to bisect the z-axis vector. The y-axis unit vector was defined as a cross product of the X and Z vectors. All joint angles were converted to complementary angles, as in the modeled papers [26–29]. Kinematic data were gap-filled, filtered, and smoothed using Butterworth filtering. Data collected included tarsus joint angles in the X, Y, and Z planes, including peak tarsus and stifle joint flexion and extension, angular velocity, and angular acceleration.

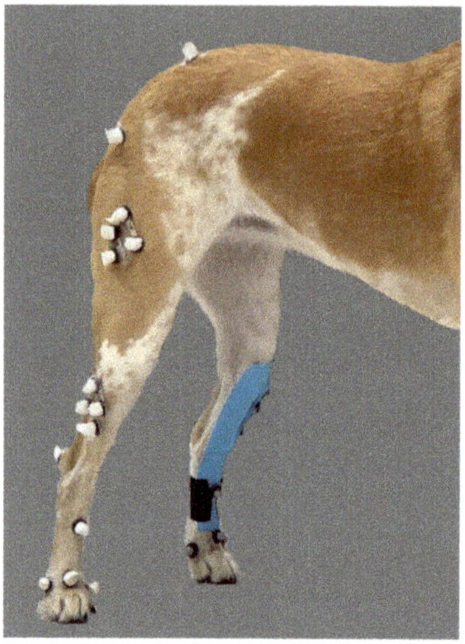

Figure 3. Instrumented dog with kinesiology tape applied.

Kinetic Data Collection:

Ground-reaction forces were determined using a force platform (AMTI OR6-6, Watertown, MA, USA) and quadruped software (Acquire version 7.33, Vicon, Centennial, CO, USA). Dogs were walked and trotted over the platform between 0.7 and 1.2 m/s and 1.7 and 2.1 m/s, respectively, with an acceleration of ± 0.5 m/s^2. Speed and acceleration were measured using five photoelectric cells mounted 50 cm apart at a height of 58 cm within the testing space. Trials were included for analysis if ipsilateral forelimb and hindlimb strikes occurred, velocity and acceleration were within the described parameters, and there were no sudden changes in limb, body, or head motion. Four valid right and left ipsilateral limb strikes were obtained, and the means of the four trials were calculated for each parameter for each side. Peak vertical (Z_{Peak}), braking (Y_A), and propulsion (Y_B) forces were determined as a percentage of body weight. Visual examination of the graphical representation confirmed the validity of trials. Kinetic data were collected before tape application and 5 min and 2 h after application.

Kinetic and Kinematic Data Analysis:

A three-way repeated measures ANOVA (SAS, Cary, NC, USA) was performed on ground reaction forces, angular acceleration, angular velocity, and maximum and minimum joint angles of the tarsus and stifle. Comparisons between the taped and untaped limbs, as well as comparisons between exercises (walking, trotting, and cavaletti poles), and exercise × time interaction were performed.

3. Results

3.1. Participants

A total of 10 dogs participated in the study. Breeds included German Shepherd, Standard Poodle, Golden Retriever Mix, Boxer mix, Great Dane, Weimaraner mix, English

Retriever Spaniel, and mixed breed dogs. The average age of participants was 6.2 years (range 2 to 10), and the average weight was 25.8 kg (range 17.4–40.4). The body condition score of dogs was 4–6 out of 10. Four dogs had the tape applied to the left tarsus, and six dogs had the tape applied to the right tarsus.

All dogs had short-to-medium length hair; dogs with medium-length hair were clipped for marker application. One dog (German Shepherd) experienced minor dermatitis from clipping. It was treated with topical dilute chlorohexidine and resolved in 5 days without complication. No other complications occurred.

3.2. Kinesiology Tape Longevity

The kinesiology tape either fell off or required additional support pieces applied within 2 h (Figure 4). In most dogs, the tape loosened within 15 min. Additional support pieces were placed perpendicularly on the limb depending on the site of tape failure (Figure 5). Failure occurred most commonly at the tarsocrural joint, likely due to the high motion and angularity in this area. Other sites of failure included the proximal and distal ends of the tape.

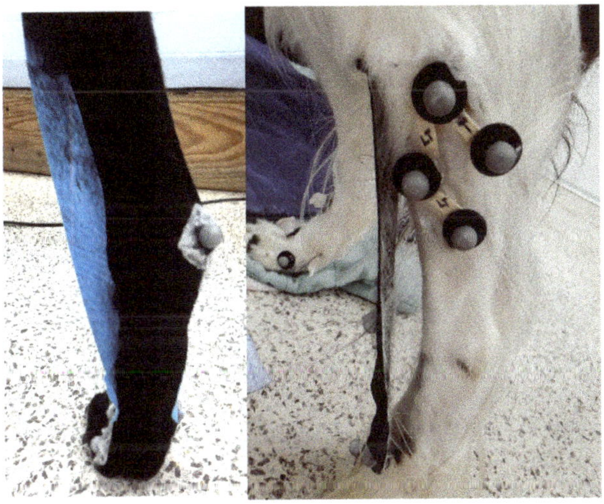

Figure 4. Kinesiology tape failure over the tarsocrural joints.

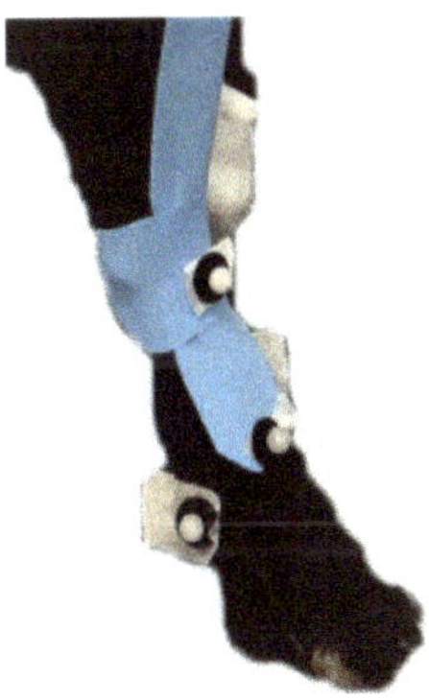

Figure 5. Additional tape added perpendicular to the original tape to hold it in place. Each subject required this modification by the 2-h test mark or earlier.

3.3. Kinematic Gait Evaluation

There were no significant differences among any of the comparisons, except for significant changes in angular acceleration and angular velocity of joint motion and tarsal and stifle flexion among exercises (Tables 2 and 3, Figures 6 and 7). Flexion was greatest with cavaletti rail walking, followed by trotting and walking. The presence of kinesiology tape application had no effect on altering the measured variables during a particular exercise, however.

Table 2. Summary of three-way repeated measures ANOVA of tarsal kinematic measurements.

Tarsus Kinematic Variables (p Values)				
Comparison	**Angular Acceleration**	**Angular Velocity**	**Maximum Flexion**	**Minimum Flexion**
Affected vs. Unaffected Leg	0.5042	0.5418	0.6946	0.7511
Exercises	0.0004	0.0001	0.1554	<0.0001
Leg vs. Exercises	0.7955	0.4757	0.3118	0.4251
Time	0.7509	0.9452	0.9960	0.3644
Leg vs. Time	0.9848	0.9041	0.5699	0.8166
Exercises vs. Time	0.0861	0.3684	0.3258	0.0596
Leg vs. Exercises vs. Time	0.2671	0.5127	0.8765	0.9506

Table 3. Summary of three-way repeated measures ANOVA of stifle kinematic measurements.

Stifle Kinematic Variables (p Values)				
Comparison	**Angular Acceleration**	**Angular Velocity**	**Maximum Flexion**	**Minimum Flexion**
Affected vs. Unaffected Leg	0.2613	0.1333	0.6581	0.9445
Exercises	<0.0001	<0.0001	0.0885	<0.0001
Leg vs. Exercises	0.3807	0.1368	0.7262	0.9023
Time	0.3194	0.2521	0.9675	0.3826
Leg vs. Time	0.6637	0.0665	0.0747	0.8477
Exercises vs. Time	0.027	0.5549	0.2693	0.3152
Leg vs. Exercises vs. Time	0.2032	0.534	0.1465	0.2343

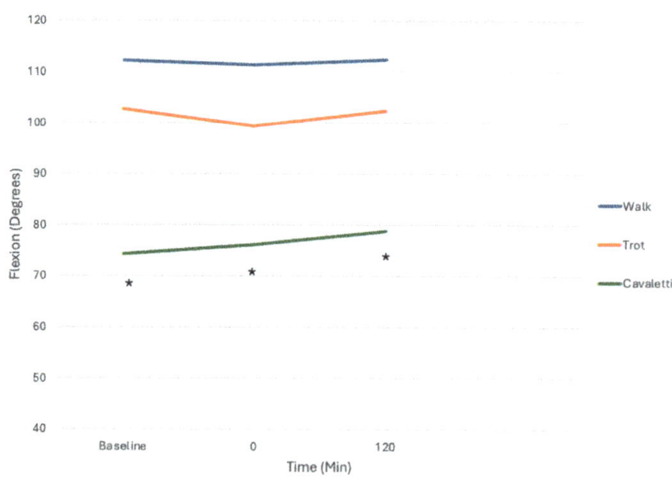

Figure 6. Stifle flexion at a walk, trot, and stepping over cavaletti rails before and after kinesiology tape application. Flexion was significantly greater with stepping over cavaletti compared with walking and trotting, but there were no differences with kinesiology tape application (* indicates significance between walking vs. cavaletti and trotting vs. cavaletti).

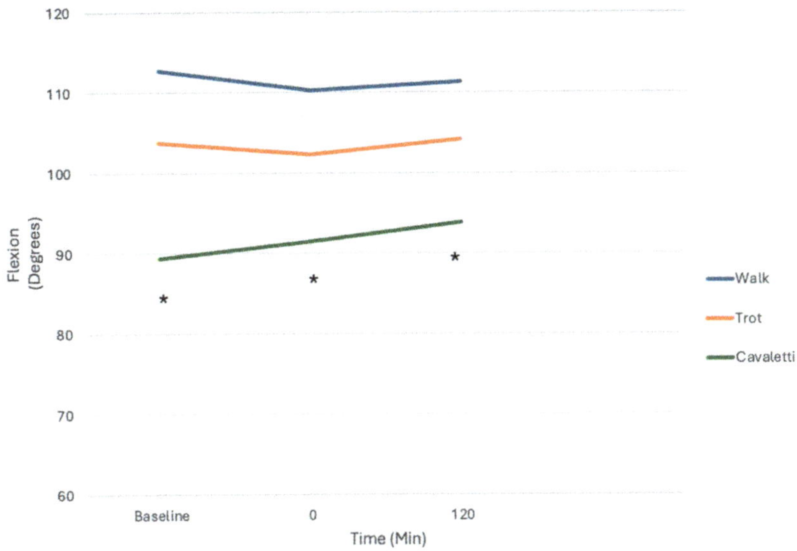

Figure 7. Tarsal flexion at a walk, trot, and stepping over cavaletti rails before and after kinesiology tape application. Flexion was significantly greater for stepping over cavaletti rails compared with walking and trotting, but there were no differences with kinesiology tape application (* indicates significance between walking vs. cavaletti and trotting vs. cavaletti).

3.4. Kinetic Gait Evaluation

Three-way repeated measures of ANOVA showed no statistical difference between the taped and untaped limbs between limbs, exercise × time, or other comparisons listed in Table 4. There were differences in mean ground reaction forces regarding exercises, with Z_{Peak} and Y_{APeak} being greater with trotting (Figure 8).

Table 4. Summary of three-way repeated measures ANOVA of ground-reaction force variables.

	Kinetic Variables (*p* Values)		
Comparison	**Peak Vertical Force (Z_{Peak})**	**Propulsion (Y_{APeak})**	**Braking (Y_{BPeak})**
Affected vs. Unaffected Leg	0.7838	0.5418	0.4610
Exercises	<0.0001	0.0002	0.2631
Leg × Exercises	0.8394	0.9706	0.9144
Time	0.7132	0.2127	0.3422
Leg vs. Time	0.4063	0.3110	0.8742
Exercises × Time	0.0514	0.5281	0.9687
Leg × Exercises × Time	0.9915	0.7742	0.2417

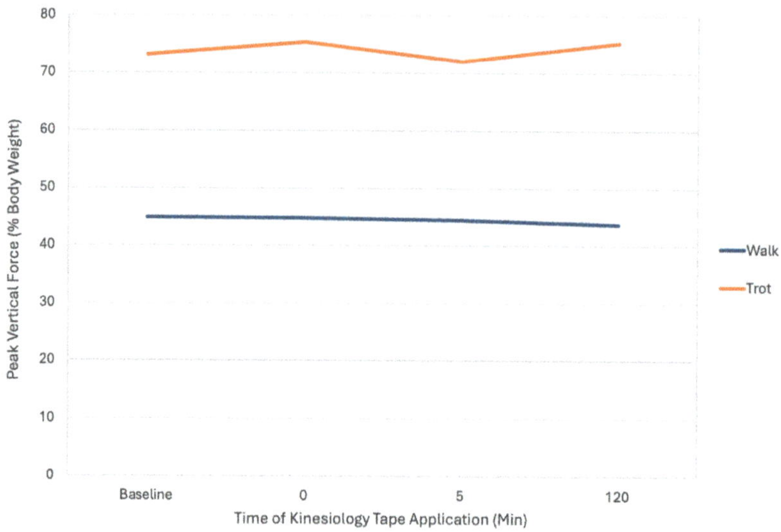

Figure 8. Peak vertical force of dogs walking and trotting before and after kinesiology tape application. Peak vertical force was significantly greater with trotting vs walking, but there were no differences with kinesiology tape application.

4. Discussion

Based on the results reported here, the application of kinesiology tape to the canine tarsus had no significant effect on tarsal or stifle joints regarding gait or performance of selected exercises. The tape had no effect on kinematic joint variables in the tarsus or stifle, similar to the effect of kinesiology taping of the forelimb in horses [23], nor any effect on measured ground reaction forces at the walk or trot, unlike a study of tape application to equine fetlocks [22]. Therefore, our original hypotheses postulating increased range of motion and decreased peak vertical force were rejected.

Studies in people have also found little effect of kinesiology tape on joint motion [30]. However, improvement in muscle strength has been reported in people undergoing kinesiology taping for knee osteoarthritis, muscle fatigue, and shoulder function during performance activities [14,31,32], while other studies have shown no effect [33–35]. Although improvement of joint proprioception has been a suggested benefit of kinesiology taping [36,37] and may be beneficial in those with poor proprioception [38], this may not occur in normal patients [39–41].

A secondary aim of this study was to evaluate kinesiology tape longevity based on the manufacturer's claims that it can be worn for 5 days. The tape was applied per the manufacturer's instructions. In this study, the kinesiology tape did not adhere to the limb as long as expected. All dogs required tape modification within 2 h of application. However, this may be due to the high mobility of the canine tarsus and may not reflect inherent properties of the tape itself or the tape application protocol. It is possible the tape may perform better on other joints or on dogs that are not performing exercises with great joint motion after tape application. It is also unknown if the additional pieces added to maintain tape position (added perpendicularly to the original piece) altered joint motion. Further studies using tape on other joints may produce different results and provide more longevity of the tape.

Based on the results of the study presented here, kinesiology tape does not alter gait or performance of the tested exercises in clinically normal dogs. The tape did not affect any of the measured parameters when compared to the untaped contralateral limb. This is similar to the effect of kinesiology taping of the forelimb in horses and in back flexion and extension, which found no effect on forelimb joint or back kinematics [9,23]. However, the lack of effect of kinesiology taping on ground-reaction forces in dogs differs from a study of tape application to equine fetlocks, which found decreased ground reaction forces [22]. Although we found no effects in our study of normal dogs, studies of kinesiology taping effects in dogs with pathology, such as osteoarthritis, should be considered [42].

Studies evaluating tape application methods may also be warranted. Although some studies have suggested that tape tension and the direction of tape application may make a difference [43–45], others have shown that the direction of the tape may not make a difference regarding athletic performance [46–48]. Changes in application, such as clipping the hair over the intended area of all study dogs, may help increase contact between the adhesive and skin. Also, tapes from different manufacturers exhibit different characteristics regarding adherence [49–51].

Limitations in this study include a small sample size and difficulties in the application of the kinesiology tape and tape longevity. Skin movement about markers is also a limitation of the kinematic evaluation of gait. However, the techniques used in this study allowed noninvasive collection of data and used no radiation equipment that has been used in other studies, such as fluoroscopy, to limit radiation exposure. In addition, the use of the virtual markers used in the study reported here has been shown to reduce the effects of marker motion artifact. Each dog served as its own untaped control, making any differences in joint motion a result of kinesiology tape application consistent.

In our original hypothesis, we believed that dogs would have altered tarsal joint movement, and we also believed that this effect would be less as the dogs became accustomed to wearing the tape. That was the logic behind repeated evaluation over 2 h or longer. However, because there were no immediate differences in joint motion right after tape application, along with difficulties in maintaining tape in position, we were unable to evaluate the dogs longer than 2 h after tape application.

To our knowledge, this is the first study to evaluate the effect of kinesiology tape on gait characteristics in normal dogs. Our study of the effects of kinesiology tape on dogs is important to further understand any impacts on clinical patients. Kinesiology tape is increasingly used in veterinary practice, likely because of its impact on human sports medicine. Systematic scientific studies are warranted to understand the true effectiveness of the tape. Future research to evaluate the use of kinesiology tape on other joints, such as the stifle, coxofemoral, and cubital joints, as well as the spinal column, is warranted in clinically normal dogs and dogs with various orthopedic or neurologic conditions.

5. Conclusions

In conclusion, kinesiology tape applied to the dorsal canine tarsus had no significant effect on gait or the completion of selected exercises. The tape had no effect on measured kinematic or kinetic variables. The tape, as applied according to the manufacturer's instructions, did not last beyond 2 h. The tape application protocol should be revisited if subjects are asked to perform rehabilitation exercises over a period of time. Additional studies evaluating tape application, tape longevity, and tape effects on other joints during motion are also warranted.

Author Contributions: Conceptualization, D.L.M.; methodology, R.C.N., and D.L.M.; formal analysis, R.C.N.; investigation, R.C.N., L.M.S., N.H.M., K.J., and D.L.M.; data curation, R.C.N.; writing—original draft preparation, R.C.N.; writing—review and editing, R.C.N., L.M.S., N.H.M., K.J., and D.L.M.; supervision, D.L.M.; funding acquisition, D.L.M. All authors have read and agreed to the published version of the manuscript.

Funding: Funding for this study was provided by the Veterinary Orthopedic Laboratory at the University of Tennessee College of Veterinary Medicine and Acree Endowed Chair funds. All funding was internally sourced. Authors do not have any financial or personal relationships with Vicon or RockTape that influenced or provided bias to the background of the research.

Institutional Review Board Statement: The study was conducted according to the guidelines of the and approved by the Institutional Review Board of The University of Tennessee (protocol code 2766-0420).

Informed Consent Statement: Informed consent was obtained from all subjects involved in the study.

Data Availability Statement: Data available on request.

Acknowledgments: The authors would like to thank all owners and their dogs involved in the study. Mike O'Neil of the University of Tennessee statistics department, Ann Viera, and the library team at the University of Tennessee College of Veterinary Medicine for their aid in the literature search.

Conflicts of Interest: The authors declare no conflicts of interest.

References

1. Banerjee, G.; Briggs, M.; Johnson, M.I. The Immediate Effects of Kinesiology Taping on Cutaneous Blood Flow in Healthy Humans under Resting Conditions: A Randomised Controlled Repeated-Measures Laboratory Study. *PLoS ONE* **2020**, *15*, e0229386. [CrossRef] [PubMed]
2. Chan, M.C.E.; Wee, J.W.J.; Lim, M.H. Does Kinesiology Taping Improve the Early Postoperative Outcomes in Anterior Cruciate Ligament Reconstruction? A Randomized Controlled Study. *Clin. J. Sport Med.* **2017**, *27*, 260–265. [CrossRef]
3. Kaltron, A.; Bar-Sela, S. A Systematic Review of the Effectiveness of Kinesio Taping®—Fact or Fashion? *Eur. J. Phys. Rehabil. Med.* **2013**, *49*, 699–709.
4. Lee, K.; Yi, C.; Lee, S. The Effects of Kinesiology Taping Therapy on Degenerative Knee Arthritis Patients' Pain, Function and Joint Range of Motion. *J. Phys. Ther. Sci.* **2016**, *28*, 63–66. [CrossRef]
5. Mao, H.Y.; Hu, M.T.; Yen, Y.Y.; Lan, S.J.; Lee, S. Da Kinesio Taping Relieves Pain and Improves Isokinetic Not Isometric Muscle Strength in Patients with Knee Osteoarthritis—A Systematic Review and Meta-Analysis. *Int. J. Environ. Res. Public Health* **2021**, *18*, 10440. [CrossRef] [PubMed]
6. Lim, E.C.W.; Tay, M.G.X. Kinesio Taping in Musculoskeletal Pain and Disability That Lasts for More than 4 Weeks: Is It Time to Peel off the Tape and Throw It out with the Sweat? A Systematic Review with Meta-Analysis Focused on Pain and Also Methods of Tape Application. *Br. J. Sports Med.* **2015**, *49*, 1558–1566. [CrossRef]
7. Montalvo, A.M.; Cara, E.L.; Myer, G.D. Effect of Kinesiology Taping on Pain in Individuals with Musculoskeletal Injuries: Systematic Review and Meta-Analysis. *Physician Sportsmed.* **2014**, *42*, 48–57. [CrossRef]
8. Parreira, P.d.C.S.; Costa, L.d.C.M.; Junior, L.C.H.; Lopes, A.D.; Costa, L.O.P. Current Evidence Does Not Support the Use of Kinesio Taping in Clinical Practice: A Systematic Review. *J. Physiother.* **2014**, *60*, 31–39. [CrossRef]
9. Ericson, C.; Stenfeldt, P.; Hardeman, A.; Jacobson, I. The Effect of Kinesiotape on Flexion-extension of the Thoracolumbar Back in Horses at Trot. *Animals* **2020**, *10*, 301. [CrossRef]
10. Drouin, J.L.; Mcalpine, C.T.; Primak, K.A.; Kin, B.A.; Kissel, J.; Drouin, J. The Effects of Kinesiotape on Athletic-Based Performance Outcomes in Healthy, Active Individuals: A Literature Synthesis Private Practice. *J. Can. Chiropr. Assoc.* **2013**, *57*, 356–365.
11. Csapo, R.; Alegre, L.M. Effects of Kinesio® Taping on Skeletal Muscle Strength-A Meta-Analysis of Current Evidence. *J. Sci. Med. Sport* **2015**, *18*, 450–456. [CrossRef] [PubMed]
12. de Freitas, F.S.; Brown, L.E.; Gomes, W.A.; Behm, D.G.; Marchetti, P.H. No Effect of Kinesiology Tape on Passive Tension, Strength or Quadriceps Muscle Activation or during Maximal Voluntary Isometric Contraction in Resistance Trained Men. *Int. J. Sports Phys. Ther.* **2018**, *13*, 661–667. [CrossRef]
13. Farquharson, C.; Greig, M. Kinesiology Tape Mediates Soccer-Simulated and Local Peroneal Fatigue in Soccer Players. *Res. Sports Med.* **2017**, *25*, 313–321. [CrossRef]
14. Konishi, Y. Tactile Stimulation with Kinesiology Tape Alleviates Muscle Weakness Attributable to Attenuation of Ia Afferents. *J. Sci. Med. Sport* **2013**, *16*, 45–48. [CrossRef]
15. Lau, K.K.L.; Cheng, K.C.C. Effectiveness of Taping on Functional Performance in Elite Athletes: A Systematic Review. *J. Biomech.* **2019**, *90*, 16–23. [CrossRef]
16. Fong, S.S.M.; Tam, Y.T.; Macfarlane, D.J.; Ng, S.S.M.; Bae, Y.H.; Chan, E.W.Y.; Guo, X. Core Muscle Activity during TRX Suspension Exercises with and without Kinesiology Taping in Adults with Chronic Low Back Pain: Implications for Rehabilitation. *Evid.-Based Complement. Altern. Med.* **2015**, *2015*, 910168. [CrossRef] [PubMed]

17. Poon, K.Y.; Li, S.M.; Roper, M.G.; Wong, M.K.M.; Wong, O.; Cheung, R.T.H. Kinesiology Tape Does Not Facilitate Muscle Performance: A Deceptive Controlled Trial. *Man. Ther.* **2015**, *20*, 130–133. [CrossRef] [PubMed]
18. Nunes, G.S.; Vargas, V.Z.; Wageck, B.; dos Santos Hauphental, D.P.; da Luz, C.M.; de Noronha, M. Kinesio Taping Does Not Decrease Swelling in Acute, Lateral Ankle Sprain of Athletes: A Randomised Trial. *J. Physiother.* **2015**, *61*, 28–33. [CrossRef]
19. Gabriel, E.H.; Sibert, N.C.; Powden, C.J. Kinesiology Therapeutic Taping Does Not Reduce Swelling Caused by an Acute Ankle Sprain: A Critically Appraised Topic. *J. Sport Rehabil.* **2022**, *31*, 630–634. [CrossRef]
20. Alexander, J.; Hobbs, S.J.; May, K.; Northrop, A.; Brigden, C.; Selfe, J. Postural Characteristics of Female Dressage Riders Using 3D Motion Analysis and the Effects of an Athletic Taping Technique: A Randomised Control Trial. *Phys. Ther. Sport* **2015**, *16*, 154–161. [CrossRef]
21. Molle, S. Kinesio Taping Fundalmentals for the Equine Athlete. *Vet. Clin. Equine* **2016**, *32*, 103–113. [CrossRef] [PubMed]
22. Ramón, T.; Prades, M.; Armengou, L.; Lanovaz, J.L.; Mullineaux, D.R.; Clayton, H.M. Effects of Athletic Taping of the Fetlock on Distal Limb Mechanics. *Equine Vet. J.* **2004**, *36*, 764–768. [CrossRef]
23. Zellner, A.; Bockstahler, B.; Peham, C. The Effects of Kinesio Taping on the Trajectory of the Forelimb and the Muscle Activity of the Musculus Brachiocephalicus and the Musculus Extensor Carpi Radialis in Horses. *PLoS ONE* **2017**, *12*, e0186371. [CrossRef] [PubMed]
24. Biau, S.; Burgaud, I. Application of Kinesiology Taping to Equine Abdominal Musculature in a Tension Frame for Muscle Facilitation Increases Longitudinal Activity at the Trot. *Equine Vet. J.* **2022**, *54*, 973–978. [CrossRef]
25. Fu, Y.C.; Torres, B.T.; Budsberg, S.C. Evaluation of a Three-Dimensional Kinematic Model for Canine Gait Analysis. *Am. J. Vet. Res.* **2010**, *71*, 1118–1122. [CrossRef]
26. Torres, B.T.; Fu, Y.C.; Sandberg, G.S.; Budsberg, S.C. Pelvic Limb Kinematics in the Dog with and without a Stifle Orthosis. *Vet. Surg.* **2017**, *46*, 642–652. [CrossRef] [PubMed]
27. Sandberg, G.S.; Torres, B.T.; Budsberg, S.C. Review of Kinematic Analysis in Dogs. *Vet. Surg.* **2020**, *49*, 1088–1098. [CrossRef]
28. Torres, B.T.; Gilbert, P.J.; Reynolds, L.R.; Fu, Y.C.; Navik, J.A.; Sornborger, A.; Budsberg, S.C. The Effect of Examiner Variability on Multiple Canine Stifle Kinematic Gait Collections in a 3-Dimensional Model. *Vet. Surg.* **2015**, *44*, 581–587. [CrossRef]
29. Torres, B.T.; Punke, J.P.; Fu, Y.C.; Navik, J.A.; Speas, A.L.; Sornborger, A.; Budsberg, S.C. Comparison of Canine Stifle Kinematic Data Collected with Three Different Targeting Models. *Vet. Surg.* **2010**, *39*, 504–512. [CrossRef]
30. Alam, S.; Malhotra, D.; Munjal, J.; Chachra, A. Immediate Effect of Kinesio Taping on Shoulder Muscle Strength and Range of Motion in Healthy Individuals: A Randomised Trial. *Hong Kong Physiother. J.* **2015**, *33*, 80–88. [CrossRef]
31. Angelopoulos, P.; Mylonas, K.; Tsepis, E.; Billis, E.; Vaitsis, N.; Fousekis, K. The Effects of Instrument-Assisted Soft Tissue Mobilization, Tissue Flossing, and Kinesiology Taping on Shoulder Functional Capacities in Amateur Athletes. *J. Sport Rehabil.* **2021**, *30*, 1028–1037. [CrossRef] [PubMed]
32. Altaş, E.U.; Günay Uçurum, S.; Özer Kaya, D. Acute Effect of Kinesiology Taping on Muscle Strength, Tissue Temperature, Balance, and Mobility in Female Patients with Osteoarthritis of the Knee. *Somatosens. Mot. Res.* **2021**, *38*, 48–53. [CrossRef]
33. Cavaleri, R.; Thapa, T.; Beckenkamp, P.R.; Chipchase, L.S. The Influence of Kinesiology Tape Colour on Performance and Corticomotor Activity in Healthy Adults: A Randomised Crossover Controlled Trial. *BMC Sports Sci. Med. Rehabil.* **2018**, *10*, 17. [CrossRef] [PubMed]
34. Cai, C.; Au, I.P.H.; An, W.; Cheung, R.T.H. Facilitatory and Inhibitory Effects of Kinesio Tape: Fact or Fad? *J. Sci. Med. Sport.* **2016**, *19*, 109–112. [CrossRef] [PubMed]
35. Halski, T.; Dymarek, R.; Ptaszkowski, K.; Słupska, L.; Rajfur, K.; Rajfur, J.; Pasternok, M.; Smykla, A.; Taradaj, J. Kinesiology Taping Does Not Modify Electromyographic Activity or Muscle Flexibility of Quadriceps Femoris Muscle: A Randomized, Placebo-Controlled Pilot Study in Healthy Volleyball Players. *Med. Sci. Monit.* **2015**, *21*, 2232–2239. [CrossRef]
36. Hosp, S.; Folie, R.; Csapo, R.; Hasler, M.; Nachbauer, W. Eccentric Exercise, Kinesiology Tape, and Balance in Healthy Men. *J. Athl. Train.* **2017**, *52*, 636–642. [CrossRef]
37. Hosp, S.; Csapo, R.; Heinrich, D.; Hasler, M.; Nachbauer, W. Does Kinesiology Tape Counter Exercise-Related Impairments of Balance in the Elderly? *Gait Posture* **2018**, *62*, 167–172. [CrossRef]
38. Hosp, S.; Bottoni, G.; Heinrich, D.; Kofler, P.; Hasler, M.; Nachbauer, W. A Pilot Study of the Effect of Kinesiology Tape on Knee Proprioception after Physical Activity in Healthy Women. *J. Sci. Med. Sport.* **2015**, *18*, 709–713. [CrossRef]
39. Blein-Ibáñez, Á.; Molina-Rueda, F.; Bebea-Zamorano, F.N.; López-Román, A.; Navarro-López, V.; Alguacil-Diego, I.M. Kinesiology Tape versus Non-Standardized Tape in the Treatment of Balance in Non-Operated Anterior Cruciate Ligament Rupture Patients: A Randomized Controlled Trial. *Somatosens. Mot. Res.* **2023**, *41*, 222–229. [CrossRef]
40. Rogers, M.E.; Jimoh, J.M.; Zhang, Q.; Patterson, J.A.; Bomgardner, R.; Rogers, N.L. *Effects of Kinesiology Tape on Balance*; Proc of American College of Sports Medicine: Boston, MA, USA, 2016; p. 142.
41. Bailey, D.; Firth, P. Does Kinesiology Taping of the Ankles Affect Proprioceptive Control in Professional Football (Soccer) Players? *Phys. Ther. Sport* **2017**, *25*, 94–98. [CrossRef]

42. Park, J.S.; Yoon, T.; Lee, S.H.; Hwang, N.K.; Lee, J.H.; Jung, Y.J.; Lee, G. Immediate Effects of Kinesiology Tape on the Pain and Gait Function in Older Adults with Knee Osteoarthritis. *Medicine* **2019**, *98*, e17880. [CrossRef] [PubMed]
43. Lopes, M.; Torres, R.; Romão, D.; Dias, M.; Valério, S.; Espejo-Antúnez, L.; Costa, R.; Ribeiro, F. Kinesiology Tape Increases Muscle Tone, Stiffness, and Elasticity: Effects of the Direction of Tape Application. *J. Bodyw. Mov. Ther.* **2022**, *30*, 176–180. [CrossRef] [PubMed]
44. Magalhães, I.; Bottaro, M.; Mezzarane, R.A.; Neto, F.R.; Rodrigues, B.A.; Ferreira-Júnior, J.B.; Carregaro, R.L. Kinesiotaping Enhances the Rate of Force Development but Not the Neuromuscular Efficiency of Physically Active Young Men. *J. Electromyogr. Kinesiol.* **2016**, *28*, 123–129. [CrossRef] [PubMed]
45. Naugle, K.E.; Hackett, J.; Aqeel, D.; Naugle, K.M. Effect of Different Kinesio Tape Tensions on Experimentally-Induced Thermal and Muscle Pain in Healthy Adults. *PLoS ONE* **2021**, *16*, e0259433. [CrossRef]
46. Choi, I.R.; Lee, J.H. Effect of Kinesiology Tape Application Direction on Quadriceps Strength. *Medicine* **2018**, *97*, e11038. [CrossRef]
47. Choi, I.R.; Lee, J.H. Effects of the Direction of Kinesiology Tape Application on the Delayed Onset of Quadriceps Muscle Fatigue in Athletes. *Isokinet. Exerc. Sci.* **2019**, *27*, 235–240. [CrossRef]
48. Dolphin, M.; Brooks, G.; Calancie, B.; Rufa, A. Does the Direction of Kinesiology Tape Application Influence Muscle Activation in Asymptomatic Individuals? *Int. J. Sports Phys. Ther.* **2021**, *16*, 135–144. [CrossRef]
49. Matheus, J.P.C.; Zille, R.R.; Gomide Matheus, L.B.; Lemos, T.V.; Carregaro, R.L.; Shimano, A.C. Comparison of the Mechanical Properties of Therapeutic Elastic Tapes Used in Sports and Clinical Practice. *Phys. Ther. Sport* **2017**, *24*, 74–78. [CrossRef]
50. Selva, F.; Pardo, A.; Aguado, X.; Montava, I.; Gil-Santos, L.; Barrios, C. A Study of Reproducibility of Kinesiology Tape Applications: Review, Reliability and Validity. *BMC Musculoskelet. Disord.* **2019**, *20*, 153. [CrossRef]
51. Tunakova, V.; Tunak, M.; Mullerova, J.; Kolinova, M.; Bittner, V. Material, Structure, Chosen Mechanical and Comfort Properties of Kinesiology Tape. *J. Text. Inst.* **2017**, *108*, 2132–2146. [CrossRef]

MDPI AG
Grosspeteranlage 5
4052 Basel
Switzerland
Tel.: +41 61 683 77 34

Veterinary Sciences Editorial Office
E-mail: vetsci@mdpi.com
www.mdpi.com/journal/vetsci

www.ingramcontent.com/pod-product-compliance
Lightning Source LLC
Chambersburg PA
CBHW061628210926
51850CB00009B/2472